FIRE
IN
NATURE

A FIRE ACTIVIST'S GUIDE

Written and Compiled by Ed Komarek

SHOESTRING PUBLISHING
Copyright by Ed Komarek
3301 Hwy 93 South, Cairo Georgia, 39828

ISBN: (13: 978-1499159226)

Fire in Nature a Fire Activist's Guide [1] is the second in a series of books written by Ed Komarek to summarize the knowledge and wisdom gained during his lifetime so that others may build on his work as he has built on the good works of others during his lifetime. The first book UFOs Exopolitics and the New World Disorder is available to be read for free on its website or downloaded as a PDF. [2]

The hard copy and the Kindle edition of Fire in Nature can be purchased from Amazon with wholesale copies available from Createspace Publishing. For at least a limited time Ed would like to offer print copies at cost, plus shipping and handling, to public and private public education organizations like Tall Timbers Inc., Gulf Specimen, Birdsong Nature Center, The Nature Conservancy, and the Prescribed Fire Training Center. Contact: edkomarek@yahoo.com

ACKNOWLEDGEMENTS

Clinton Bailey has created the excellent book design and cover for this book which is very much appreciated. www.theartistree.com.au I would like to thank my longtime friend Thomas Reeves for helping me edit this book. I wish to thank Tall Timbers Research Inc. for the use of their extensive fire library and other help, as well as other sources who wish to remain anonymous, who have been very helpful in helping me get up to speed on modern fire management practices.

NOTE TO READERS

The source material for this book is referenced using footnote links for those reading this book free on the Internet at its website URL. http://fireinnature.weebly.com/ For those not reading the book on the Internet, a list of all the footnoted sources with URLs are listed at the end of the book. Ed Komarek can be contacted on Facebook message or email at: edkomarek@yahoo.com

DEDICATION

This book is dedicated to my ecological mentors and fellow fire activists. From the time I could walk, I was mentored in ecology and specifically fire ecology by my parents Ed and Betty Komarek, Herb Stoddard, and my Uncle Roy Komarek. The mentoring started on Birdsong Plantation, Tall Timbers and Greenwood Plantation. During early family travels, the mentoring continued onto the public lands of the United States, Canada and Mexico.

I offer thanks to the other founders of Tall Timbers Research Station that include Leon Neel and Henry Beadel. Tall Timbers was later expanded to be called Tall Timbers Research Station & Land Conservancy after the founders had died or had to leave. Thanks also to Robert Crawford and Wilson Baker, early loyal employees of Tall Timbers that have been supportive to this day working to keep the founder's vision of continued fire research and activism alive.

I must also include John Hay Whitney who owned Greenwood Plantation and who's generosity allowed my father's research to be part of his regular day job at Greenwood. The reason my Uncle and my Dad were able to devote so much free time and personal resources to build Tall Timbers in its early years, was because of John Whitney's dedication to   Ed, Roy, Herb, Leon and Henry's struggle against fire exclusion in light fire ecosystems.

As the focus shifted at Tall Timbers away from its leading role in early fire research and activism and into land conservation, other popular writers like Stephen Pyne and organizations like the Nature Conservancy and the Prescribed Fire Training Centers have taken on new leadership roles in the continuing struggle to put light fire back into light fire ecosystems.

TABLE OF CONTENTS

PREFACE

I was raised in a family of early ecologists who mentored me in ecology and fire ecology and hoped that I would follow in their footsteps. I went to college at the University of Alaska in Fairbanks majoring in wildlife management. I was disappointed after a year in a local junior college and two years at the University of Alaska that I was still struggling under a load of classes with little do with wildlife management.

In three years I had taken not a single course on wildlife management. I had, however, been working for Alaska Department of Fish and Game in the summers and even co-authored the first scientific paper on the Fishes of the North Slope of Alaska in the early 1970s. I was mainly responsible for the field work for this paper.

Little did I realize when I first began college that my life's direction would soon change and I would develop a lifelong interest in the paranormal. A chance encounter with a small metaphysical bookstore in downtown Fairbanks amongst the bars and shops led to an existential crisis in my life that resulted in me quitting college to try to understand myself and figure out what was life really about. I came out of this crisis a much different and wiser person several years later.

My first book UFOs Exopolitics and the New World Disorder is the first in a series of books on diverse subjects that I am writing and publishing. Fire in Nature, A Fire Activist's Guide, is the second book in this series. I am writing these books and making them free on their websites so that others may profit from what I have learned, just as I have built my life on the good works of others. There may be some fire managers, scientists, and others who may think I wear a tinfoil hat for investigating and writing about UFOs. However, I believe both books should stand on their respective merits.

These books, even while covering very different subject material, are written in a style based on exhaustive, credible footnoted source material as will be future books covering international politics and spiritual development. It would be a mistake to reject out of hand any one of these books simply because of a topic covered in another book. I am reminded of those respectable French "scientists" hundreds of years ago who refused to even look through a newly built telescope because, it obviously had to be some kind of a trick.

My parents and their ecological associates and friends were of course disappointed that I would not be following an ecological career, but they understood that each person must find their own way in life. My mother and I ended up sharing many of the same interests over the years, with my Dad and Uncle not quite able to figure out just what we were about. ☺

However, I think my family would have been delighted that I have written this second book Fire in Nature, A Fire Activist's guide to follow up on their unfinished business. The body of knowledge I have gained through a childhood apprenticeship on fire ecology is rare, and I would like this knowledge in the public domain.

I also share my parents and their associates distress and even outrage that so much public and private light fire ecosystems are being devastated by man caused catastrophic fire. One would have thought that their lifetime of work would have at least caused the losses to wildfire to be on the decrease but that has not been the case.

The reasons for the continued loss of millions of acres of light fire forest and grassland ecosystems to catastrophic wildfire, is complex, requiring a book to put all the pieces together. While there are many others writing on the need for prescribed fire in fragmented light fire ecosystems to simulate natural ecosystems, there are few willing to really stand up and not pull their punches.

Many land managers cannot stand up and exercise their free speech within the public land management agencies for fear of loss of their jobs or being demoted. Part of my job in this book is to charge in and break trail so to speak. I hope to give many of these good public bureaucrats and land managers pushing for fire management reform some cover to maneuver within their respective organizations.

Wildfire almost took out Yosemite National Park in 2013. The wildfire came into the Park. The Giant Sequoias are part of a light fire ecosystem going back to the Cretaceous, but are no match for catastrophic fire undergrowth and debris build up over decades of man's misguided light fire suppression.

Are we to devastate this park by wildfire and destroy these thousand year old trees, same as happened with Yellowstone National Park? Are we to consider the loss of our parks and national forests inevitable blaming climate change, fire, arson and lighting, anything but the real reason this devastation caused by our own fire suppression culture?

FORWARD

I am very pleased to see that my longtime friend Ed Komarek Jr. (Eddie) decided to write a fire ecology book to pick up where his father Ed Komarek Sr. left off when he died. Ed Sr. played a profound and critical ecological role in my life and was critical to the survival of Gulf Specimen Marine Lab. in its early years. Without his encouragement, advice and financial support, the aquarium and marine lab that exists today in Panacea, Florida could easily have fallen into oblivion.

I have been an interested party watching mostly from the sidelines on this particular fight as the founders of Tall Timbers were able to educate the public and pressure government land management agencies to put fire back into its rightful place here in the Southeastern United States. As Ed Komarek Jr. points out in this book, the same can't be said for the Western United States and Australia. I like other Americans see in the American media every summer that millions of acres of public and private lands are still being devastated by man caused catastrophic fire.

Despite warnings from fire ecologists both in the East and the West, Yellowstone National Park went up in smoke, as did large tracts near Los Alamos and Queensland Australia more recently. It is very upsetting to see continued mindless government bureaucratic fire suppression in the face of such unnecessary environmental disasters. It's so disturbing to watch these disasters repeat over and over again, year after year, in spite of the fact that research clearly shows that fire suppression itself is the main cause for the these unnatural fuel buildups. Natural, light, ecologically beneficial fires have been turned into huge unstoppable catastrophic fire monsters through improved and expanded fire suppression technologies.

I have spent most of my life fighting for local forests and wetlands in the Southeastern United States that are being destroyed by loggers, developers, corrupt politicians, incompetent bureaucrats and ivory tower academics. As I have fought my own often lonely battles against impossible odds, I can appreciate what Ed and Roy went through trying to put a stop to these man caused catastrophic fires worldwide with very limited success in the American West and in Australia.

Due to global over-population and over consumption, the world is fast running out of resources to be exploited, and the best we can do is engineer a "fighting retreat" and try to protect what fragmented forests and wetlands we possibly can from continued needless destruction by man. Ed Komarek Sr. was one of my strongest mentors, the help he provided came at a critical time, and without it it's unlikely I would have been able to survive. In 1966, after returning from Madagascar on the International Indian Ocean Expedition, I moved into a leaky shack in Panacea and started collecting marine life for schools and research laboratories. I learned to collect, maintain and ship marine specimens while trying to write books and articles to stay alive.

Ed was no ivory tower scientist, before starting Tall Timbers; he developed and directly sold seed corn to stores. He understood the difficulties I had at the time with Florida State University, the enmity some biology professors bore me because I was popularizing and commercializing science, and worse yet had no college degree. Against their advice, he used funds from the newly created Tall Timbers Research Foundation and gave me a small stipend. Much was demanded for little.

Even though I was cold, wet and tired after a day of slogging around tide flats collecting marine animals, Ed and Roy insisted that I go through my buckets and write down everything while it was fresh in my mind. I could rest later, they said, but they pressed me no harder than they pressed themselves. Ed, his wife Betty and his brother Roy worked tirelessly, building Tall Timbers Research Station, Greenwood Farms, and Bird Song Nature Center. Now that I think of it he helped my writing style enabling me to write books and articles for National Geographic, the Smithsonian, and Sports illustrated, etc.

In order to prepare museum specimens, I stayed up during the night narcotizing sea anemones and preserving them with their tentacles extended. Before long we had shelves filled with pickled sponges, sea squirts, fish and seaweed, all of which were thrown out when Ed died. Ed and Roy taught me discipline in keeping records, in preparing, cataloging and organizing museum specimens. Hopefully those records are buried somewhere in the Tall Timbers archives, they will be of use in reconstructing what the coast was like before the plague of development set in.

Long before I met Ed Komarek, I used to watch his "Rural Report" along with every farmer, fisherman or anyone else who wanted to know the weather. Charismatic is too weak a word to describe him, he just seemed to engulf you in his ideas and nonsense and sweep you along. Ed added humor and playfulness to his programming; he didn't know what it was to take his self seriously. By today's television standards, Rural Report would be considered hokey, but with a small rural based population of Thomasville-Tallahassee, it was the show to watch. Rural Report had playfulness about it.

Old timers still remember and joke about how he said he crossed corn with a pine tree. A local 4H Club built an ear of corn out of wood, a yard long, plastered kernels on it and gave it to Ed. He joked about how it would help both forestry and agriculture if they could be grown on the same plant at the same time. He got people laughing and considering the possibilities. Back then, no one had heard of genetic engineering as the science was based on creating new plants by cross pollination. Gene manipulation came later.

Like other television stations, Komarek got his information from the US Weather Service that fed a ticker tape into the studio at WCTV in Thomasville, Georgia. He liked to "cut the fool" as he called it, by cranking up his weather machine, an old Corn Sheller that cranked and rattled and turned a wheel mounted with a ridiculous grinning face of the sun. He had such charisma you almost expected to hear the patter of raindrops on the roof. Everyone was in on the joke, he got his viewers laughing.

That was in the Golden Age of television and anything went. Ed used his weather generator to teach. Standing there in his faded coveralls, flannel shirt and his hay-seed straw hat, he talked about the low pressures developing in the Gulf of Mexico, and the battle of fronts, and how they stalled and deluged rain and flooded the woods and swamps, all this before the age of satellites, doppler weather and computers.

His audience grew; people listened to his insights on the role of fire in the ecosystem, the interrelationships between rainfall, wildlife and fish. If anyone listened for more than a minute, it was clearly that behind this country bumpkin and his foolery, was a man with profound knowledge of nature, agriculture, forestry, geology and meteorology. He was ready to explore new fields, including marine biology, by having me come on television twice a month.

Viewers who were around in the sixties still break out into laughter, when they recall my first appearance, and how Ed panicked the camera man and terrorized Anna Johnson, who was WCTV's secretary, by turning my horseshoe crabs loose in the studio. You could hear her shrieks in the background as he chased her around the set with my alien monster. Later he shoved her on screen, got her to do the weather, and launched her career in television news casting.

He launched mine in promoting my books and environmental causes. I brought live sea horses, starfish, even a live sawfish. The experience enabled me to make appearances on the NBC Today Show, Good Morning America, on CBS's "To Tell the Truth", PBS documentaries and numerous other programs across the nation. Edwin Komarek was a true mentor, he empowered people. He was a generous wise man that should be remembered and given prominence for his central role in creating and directing Tall Timbers Research Station, backed up by his brother Roy, in its critical early years.

Ed brought the public into the battle against the United States Forest Service and Smokey the Bear. He lectured on fire ecology, talked about the Indians burning their land, and how fire grew quail. He fought against the stigma of burning and how it was bad. Ed and Roy insisted that I attend their fire ecology conferences, and I spent many an evening standing around a fire, meeting scientists, listening at his symposiums, learning about birds. Here I met Herb Stoddard, a shy old gentleman, who worked with birds. Herb showed me his trays of stuffed birds, and explained how their differing beaks were adapted to eat seeds, bugs, or catch prey.

Ed gave both formal and informal lectures where the knowledge just flowed out and seeped into the people around him. He took me under his wing, explaining fire ecology, his vision for Tall Timbers to a twenty two year old kid, as we banged around the South Georgia and North Florida country dirt roads. He was also a showman, tossing a match onto the dry wiregrass, starting a fire. I learned about Long Leaf Pine and wire grass, at the time I didn't know one plant from another. I saw the dramatic differences between his burned plots, versus unburned plots of ground in the woods. He pointed to the lightning strikes on trees and the burned charcoal on the bark where fires had passed through. I learned about renewal, of competition, and cooperation in the forest.

As a marine ecologist and biologist I appreciate the powerful impact the flow of nutrients from the land into the ocean is having on our wetlands and marine life. Nutrients flow down from rotting vegetation and controlled burns in the uplands and marshlands, to nourish many of the species in the ocean that end up on the menu of our local restaurants supporting both the fishing industry and tourism.

It is essential that we try to protect not just the wetland and marine ecosystem, but create protective upland buffers from logging and development, and use prescribed fire to simulate light natural fires in areas particularly sensitive as nurseries to marine life.

Ed Komarek innately knew based on his ecological understanding, that something like this was going on, in that he expanded the reach of Tall Timbers right out into the Gulf of Mexico supporting my work in its early years. This is an area of research that that deserves much study in the years to come, because if it's going on here on the Gulf Coast, is bound to be happening elsewhere in marine and land ecosystems around the world.

Note: Jack Rudloe is a marine biologist- ecologist, writer, author, and environmental activist residing in Panacea Florida just south of Tallahassee Florida. Jack is also President of Gulf Specimen Marine Lab. and received in 2014 the National Wetlands Award for his and his wife Anne's lifetime of work to save and preserve Gulf Coast wetlands. [3]

INTRODUCTION

Our universe began in a burst of cosmic fire. This catastrophic event was soon followed by the fiery ignition of burning suns that irradiated their planets with light and heat, giving rise to life on earth and most likely life elsewhere as well. Over time the earth cooled and formed a planetary crust and oceans. Scientists believe that life on earth evolved from already evolving chemical processes around fiery undersea volcanic vents, or perhaps even seeded from space. Life then spread to the surface of the oceans, where simple organisms began feeding on sunlight and ocean nutrients forming the basis of the food chain to this day.

As organisms became more complex evolving into plants and animals, they began to colonize shorelines creating microbial mats that soon spread inland diversifying into a great variety of fungi, lichens and mosses. The early moss built wetland bogs, creating a buildup of dead organic material known as peat. We can speculate that on occasion in times of drought, these bogs dried out on the early earth and burned, ignited by lighting just as they do today. In such a manner the bog, rejuvenated by fire, fills up with open water when the rains come, and the process of succession begins anew.

Over hundreds of millions of years the continental tectonic plates have drifted all over the globe. This drift has created vast super continents like Pangaea where species mingled together for a time. Some species perished while others flourished as they competed and cooperated among themselves for life's necessities. When the super continents inevitably broke apart, this sent different species on different evolutionary trajectories caused by changing environmental and genetic conditions in different locations and climate conditions.

Over thousands, millions and hundreds of millions of years the climate fluctuated extensively from very warm to very cold. The cause would be internal and external impacts like asteroid impacts, massive volcanic eruptions, continental drift and even massive global change caused by the evolution of life itself. For instance some scientists speculate that one global ice age was caused by the colonization of the land by mosses. The mosses caused atmospheric changes that plunged the whole earth into a deep freeze, causing the extinction of 90% of the animals and plants at that time.

Through it all there was always fire, just as much a force of nature effecting natural ecosystems as is climate, water, atmosphere and continental drift. It is obvious that when vegetation dies it either must eventually decompose or burn if not buried by geologic processes. In warm moist environments as in the tropics it can quickly decompose. In dryer climates where decomposition is much slower fire allows nutrients to quickly get back into the soil, reduces plant and animal diseases and pests and eliminates the mulching out of new growth.

Sometimes the vegetative fires during geologic times were diminished, as during ice ages and when oxygen levels were low. Other times fires were very prominent and frequent, when oxygen levels were high and the earth warm as in the Carboniferous and Cretaceous periods. During these periods of warm climate just about every kind of plant and animal developed adaptations to fire, becoming in one way or another fire dependent.

Most people today are unaware of the critical importance of fire in nature's ecosystems and this has led to severe environmental consequences for all life on earth including man. It is imperative that environmentalists, politicians and land managers both (public and private) understand this critical role of fire in the environment in order to adequately preserve and protect the fragmented ecosystems still remaining on earth.

Most ecosystems around the globe are so fragmented by modern man that light natural fires can no longer be allowed to burn freely and frequently as they have for hundreds of millions of years. This has resulted in huge unnatural catastrophic accumulations of fuel in forests, savannas and grasslands. Sooner or later these fuel loads will be ignited by lightning or man, forcing huge unstoppable catastrophic fires damaging to both nature and man.

According to the statistics in Wildfire Today, [4] the average number of acres devastated by wildfire in the United States lower 48, has risen steadily from a little over 2 million acres in 1990 to above 6 million acres in 2013. An article in Headwaters Economics [5] states that U.S. National wildfire fighting costs have averaged $1.8 billion annually for the past five years. Costs are set to explode to between $2.3 and $4.3 billion. They compare this figure with the Forest Service's average annual budget of $5.5 billion. Talk about creating one's own worst nightmare.

So the remedy for fire in the environment is not more fire suppression, but controlled or prescribed fire to simulate natural fire. Plant and animal life adapted and used fire for competitive advantage for at least 420 million years as evidenced in the fossil record. This has created a wide spectrum of planetary diversity from very many plants with at least some fire resistance, to a very few fire tender plant species.

During the fiery Carboniferous period of geologic history oxygen levels were much higher than they are today. These higher oxygen levels allowed dead vegetation accumulating in early forests and savannas to burn much better than today. Many species of plants like the palms evolved very fire

18

resistant trunks to protect against frequent ground fires. But these early forest plants went even further to use fire to their competitive advantage.

The palms evolved flammable fronds that when the individual fronds died and fell to the ground they burned out the competition. By becoming so fire adapted they thrived at the expense of their competitors and so became a fire dominant species spreading all across the supercontinent of Pangaea. Other palm species like the palmetto and the ferns closer to the ground formed the understory and adapted by developing flammable fronds and strong underground root systems. When frequent fires burned, killing everything close above ground, they quickly sprouted back using the energy reserves stored in their roots.

Later, pine trees evolving in the fiery Cretaceous when oxygen levels were again very high and evolved thick insulating bark and flammable needles. They followed the example of the palms using fire to burn out the competition. Later the grasses evolving in these high fire environments used the same fire adapted natural strategies as had the ferns, palms, pines and the palmetto to burn out the competition and to sprout back quickly after a fire and grazing.

In a short span of only a few days when it is warm and after a rain, a blackened burn will once again become green. The grasses sprout from their fire protected roots underground in the same manner following the example of the palmetto and the ferns. Additionally, the fresh ash with the first rain soaks into the soil giving this new growth even more vitality.

All this time while the plants were evolving in fire ecosystems in Pangaea and Gondwanaland the land animals were not standing still. They were also adapting to low intensity and occasional catastrophic fires. The best I can tell the ocean arthropods began moving onto land as they do today to colonize and feed on the detritus being washed up on beaches and in lagoons where early ocean plants were also taking root. These early arthropods evolved into the insects we have today and went from jumping, to gliding, to flying, to get around and to feed on fresh vegetation and other insects on land. In the Carboniferous the fishes followed the insects onto land and evolved into amphibians and reptiles.

I would speculate that one of the reasons that insects developed wings was to get out of the way of frequent ground fires as well as predators. They also could dig into the damp ground or crawl down in openings at the base of plants. The amphibians and reptiles had to use other strategies like fast flight, or diving into water to avoid getting burned. It was not until the dinosaurs developed wings that land animals were able to fly to avoid fire or to fly back into the blackened burn to forage for roasted insects, small reptiles and small mammals now openly exposed to view.

Most people all over the world know about the almost indestructible palmetto bug or cockroach. Well, it's an example of a very well adapted fire species that evolved in the Carboniferous among the palm trees and the palmetto. Cockroaches are able to scurry down into the ground, or fly away to avoid fire, only to quickly return to feed on the greening vegetation and any other insects that might have not been so well adapted.

I am amazed that when we simulate natural low intensity fires in fragmented ecosystems that we are burning through a mosaic of ancient and modern plant and animal species. In a layering process over hundreds of millions of years, newly evolving species of plants and animals enter the ecosystem forcing out less adaptable species. However, when new species evolve in the ecosystem, they often do not always destroy earlier plants and animals. The older species may simply be pressured into moving into new niches being created by the newer invading species.

Even the first land colonizers, bacteria, fungi, algae, lichens, and mosses, still have a place in modern ecosystems adapting to the ever changing environment. Once these early land species colonized and broke down the rocks on the early earth to create the soil for trees. We see these almost identical ancestors living in and on the soil and on the trunks and branches of the trees that displaced them where they are still exposed to fire. Even our own bodies, inside and out, have more microorganisms than cells. There are about two and a half pounds of bacteria in our gut that digest our food and turn the food into usable nutrients.

We now know that when man evolved in the grasslands and savannas of Africa, he too became fire adapted, just as had other species of plants and animals living in these environments. Plants in a rudimentary way can manipulate fire, but Man went much further. He consciously learned to use fire as a tool (a paintbrush) to paint on the living canvas in which he lived in order to improve his livelihood.

This giant step made him an artist. Painting on cave walls or on canvas was simply a natural extension of his altering the landscape in which he lived. The landscapes we think of today as natural are only so in the context that man is part of the environment. Many of the plants and animals and even whole light fire ecosystems owe their continued existence to man's fire activities.

Misguided global fire exclusion and suppression policies for the past 120+ years threaten the very light fire ecosystems that nature created over hundreds of millions of years and the ones that primitive man has been busy creating for tens of thousands, if not hundreds of thousands of years. The problem we have today all around the world is that as man moved into artificial environments and out of the natural world, he has lost the knowledge that he and fire are part of the natural order of things.

Our global catastrophic wildfire problems really started when the European colonists decimated the native fire managers through disease and war around the planet. It's no accident that heavy unnatural fuel loads began to build around the globe increasing dramatically in the 1800s and continuing into the present. As the knowledgeable native frequent fire users were wiped out or displaced into reservations in the 1800s and 1900s, they have been replaced by less competent, citified, inexperienced government agents, scientists and bureaucrats who decided to unwisely and irrationally suppress fire.

Many of these early European bureaucrats were brought up in less fire prone areas of Europe or in cities ignorant of the role of fire in the environment. They were well meaning, but only knowledgeable of the devastating city fires and the unnatural catastrophic fires of the time. They thought if wildfire is bad for the city; it must be bad for nature. In the early 1900s the rising unnatural heavy fuel loads caused by displacement of the native frequent fire managers, clear cutting and wanton economic development combined to create major catastrophic fires in forests and grasslands that even spread to cities.

With ignorance building upon ignorance, the early European land managers with public support used advanced technology to seriously suppress fire throughout the 1900s. This disastrous cultural legacy has been communicated down through generations of bureaucrats to today's public and private land managers. These land managers are discovering that no amount of technology or resources can stop the catastrophic wildfires created by such huge buildups of fuel. Those interested in the catastrophic wildfire statistics for the United States can go to the National Interagency Fire Center website. [6]

Further adding insult to injury in trying to reduce these catastrophic fires, still inexperienced and underfunded fire managers have resorted to prescribed controlled fire with limited success, allowing controlled fires to get away and burn parks like Yellowstone to the ground. This only helps reinforce and supports the views of those who support fire suppression interests that compete for public money and resources with the fledgling government fire management units.

The recent and continuing catastrophic wildfires in the Western United States, Australia and Europe, that erupt with the power and devastation of multiple atomic bombs, should provoke global public outrage. This should be especially true for those land managers who understand the nature, scope and causes for accumulating fuel loads the past 120+ years in nature's global ecosystems. Many or most catastrophic wildfires can be directly traced to the fire suppression activities of man himself. Fire suppression goes against the natural order of things where light cool fires used to sweep the forest clean. Light intensity fires are the norm in most ecosystems with natural catastrophic fires limited mostly to cold climates like Alaska and Siberia.

I think it is about time to summon the ghost of Ed Komarek Sr. my father and ecological mentor as a young boy and his fire ecologist friends and colleagues. Ed Komarek Sr. has been considered by many to be the most prominent global fire ecologist of the 20th century. He, backed up by his brother Roy Komarek, organized other fire ecologists from around the world through fire conferences north of Tallahassee, Florida.

They were instrumental in the creation and operation of Tall Timbers Research Inc. in the late 1950s, 1960s, 1970s and into the 1980s. [7] Tall Timbers was organized to be a scientific, educational and activist pro-fire bulwark against the powerful misguided propaganda and fire suppression operations of the US Forest Service and other government agencies in the United States and elsewhere around the world.

Ed had been raised in Chicago where he had been taught that fire was bad for the environment. His father had gone bankrupt in the Great Depression and he had to quit college to go to work collecting mammals in the Great Smokey Mountains. When he traveled south and met and worked for Herb Stoddard his ecological mentor, he realized that fire was an intrinsic part of the natural order and to suppress fire was a crime against nature!

Herb was one of the founders of the emerging field of ecology at the time and was a good friend of another founder Aldo Leopold. Like many of the early ecologists Ed, Roy and Herb were museum collectors who because of their collecting experience realized that plants and animals were in dynamic relationship with each other.

Ed soon realized along with Herb and a few others, who Dad fondly called mavericks (cattle who would not run with the thundering herd), that suppressing light natural fires over decades created huge fuel loads in both private and public forests and grasslands. This resulted in turning small harmless and nourishing cycles of regeneration into large catastrophic cycles of catastrophic regeneration in light fire forests, savannas and grasslands.

I think that my father would have been appalled today to see the problems, while changing for the better in the Eastern United States, have been still building in the Western United States and other parts of the world like Australia. This has happened despite a lifetime of effort at the grass roots to educate and train a new generation of pro-fire land managers on public and private land.

He would surely have been disappointed that entrenched bureaucracy, public ignorance, special interests and corrupt politics at the highest levels were still responsible for continuing and amplifying these crimes against nature and man! I would say it has certainly been the case for me as a bystander in this drama, because I can see that these catastrophic wildfires are so unnecessary.

Many catastrophic and devastating fires in the United States and around the world are the result of continuing incompetence and lack of funding at the highest levels of governments. This lack of competence, excellence and funding by national and global leadership, is in part due to inadequate oversight and accountability by the public and is caused by a general public ignorance of fire's role in nature.

In turn, this failure to hold leaders accountable has been brought about in part by the dark legacy of 120+ years of misguided fire suppression, fear and propaganda by governments, an assault on public consciousness. This even throws into question the feasibility of centralized bureaucratic control over management of public lands. Surely the public deserves better that this from their public servants and managers of the public lands in America and around the world.

All this environmental destruction and loss of lives and property is just too much for me to stand by and not try to follow up on my father's unfinished business. Ed had intended to write a popular fire book before he died, but it was not to be. So in appreciation of my father and the public need, I thought I would write this fire book.

It is my desire that once the readers have read this book that it will add to a comprehensive understanding of the role of fire in nature. I hope the citizen will understand the need to reform and decentralize the dysfunctional bureaucratic system now in place. It is my hope that citizens and leaders of all stripes and persuasions, armed with fire knowledge and native wisdom, will be able to take constructive action to better preserve and protect nature's ecosystems, man and the earth itself.

It is imperative that we better educate and inform the layperson as well as the scientist, bureaucrat and politician. We can't just study and debate fire and deteriorating fragmenting ecosystems to death, we have to become involved activists and change the culture. Good science can support culture, but it is culture that drives action and consequently reform. We the people have a damaging and dysfunctional culture of fire exclusion and suppression in global land management agencies. The current systematic deplorable situation is comprised of a mix of bad science, bad economics, bad politics and a 120 year assault on public consciousness.

It is obvious that we are dealing with an entrenched culture of failure the past 120 years in these over centralized and unwieldy government land management bureaucracies. There is a failure to adequately acknowledge past misdeeds, a failure to remedy those misdeeds and a general failure to protect nature and the public interest.

When a business is failing, hired guns are sent in to observe and then to make things right when current management can't reform itself. It's a painful process for the organizations, as people responsible for failure and excuses are fired, or repositioned in a management shakeup. Still, this is not nearly as painful as bankruptcy or the catastrophic consequences to the public and nature as is now the case.

Some will be sure to argue, well what about our successes? My answer to that question is success as measured against what? I measure our very limited success in fire management against the continued increase in fuel loads globally, resulting in massive catastrophic fire catastrophes destroying many lives, property and whole fire global ecosystems. Let's not get lost in the details and forget about the big picture!

There have been great strides made in the past 60 years in both scientific fire research and management that has greatly increased the understanding of light fire burning and the importance of native people's use of fire in nature. Problem is, we now have an ever growing multibillion dollar firefighting industrial complex that incompetently throws money and resources at the problem with no overall reduction of fuel loads. This firefighting complex is not going to roll over and play dead while its funds are diverted away in order to give it only a supporting role for general fire management operations.

Making things even worse is that there is evolving a huge national security threat due to the buildup of fuel in the western United States. The terrorists have already been urging their people to use these high accumulations of forest debris to attack our western forests and urban centers perhaps to kill thousands, maybe millions of people with just small amounts of resources.

I don't think Homeland Security has any idea what a couple of people in one or more small planes could do to very quickly under the right drought and wind conditions. I believe terrorists in a couple of hours could create firestorms with the power of multiple atomic bombs blown into, or encircling western cities. It could be like the firebombing of Tokyo.

My father and his associates warned that unless something was done, parks like Yellowstone would burn to the ground by catastrophic fire, and that is exactly what happened. I in turn warn that there is a very high probability of terrorist wildfire attacks in the United States and around the globe in the next few years, but is anybody listening?

Homeland security is already informed as to the threat as expressed in the article Homeland Security Warns of Terrorist Wildfire Attacks. [8] The article states:

> The Federal Bureau of Investigation, Department of Homeland Security and fusion centers around the country are warning that terrorists are interested in using fire as a weapon, particularly in the form of large-scale wildfires near densely populated areas. A newly released DHS report states that for more than a decade "international terrorist groups and associated individuals have expressed interest in using fire as a tactic against the Homeland to cause economic loss, fear, resource depletion, and humanitarian hardship." The report notes that the tactical use of fire as a weapon is "inexpensive and requires limited technical expertise" and "materials needed to use fire as a weapon are common and easily obtainable, making preoperational activities difficult to detect and plot disruption and apprehension challenging for law enforcement."
>
> Though law enforcement has been warning of fire as a weapon for years, the recent fervor over wildfires as a potential terrorist tactic is largely due to Inspire Magazine, a slick online publication that is reportedly produced by Al-Qaeda in the Arabian Peninsula. The most recent issue of Inspire featured multiple articles on the use of wildfire as a weapon in jihad, including a complete guide on creating an "ember bomb" that would likely have a "high failure rate when manufactured and utilized by untrained or inexperienced

personnel" according to the DHS report. The FBI also has warned about the latest issue of Inspire, which "instructs the audience to look for two necessary factors for a successful wildfire, which are dryness and high winds to help spread the fire. Specific fire conditions that are likely to spread fire quickly are Pinewood, crown-fires (where the trees and branches are close together), and steep slope fires (fire spreads faster going up a slope)." California and Montana are specifically listed in Inspire as potential targets.

Pain and suffering are necessary factors that drive cultural reform and we certainly have had enough of this pain already. The question really is: how much more pain is required before the necessary changes are made? Our collective job is to change the culture of fire suppression to one of good fire management. Ultimately, it will not be public officials, scientists and business that protect and constructively manage global fire environments and the multitudes of plant and animal species dependent on fire. It's going to take an informed, aware and activist public demanding and forcing change to do the job.

In order to begin to improve our fire management practices and reform the global land management bureaucracies, we must first understand the history of fire in vegetation that started hundreds of millions of years ago. I cover this in depth in the first chapter of this book. The second chapter is on man and fire, because the understanding of the history of fire in nature by itself is not enough. This is because starting hundreds of thousands of years ago, man's ancestors and then man himself, arose out of the natural fire environment as a fire species. He even began to alter natural landscapes with fire for a livelihood.

Once we get a grasp on this history in the first two chapters of this book, then the following chapters of the book carry the reader into the arena of present fire practices and the desperate immediate need for global land management reform. At the end of the book I present suggestions as to how to improve and decentralize our present centralized dysfunctional fire and land management bureaucracies.

Key to this process of reform and transformation I believe will be the integration of past human understanding regarding the land manager's long lasting close intimate relationship to specific parcels of land. Instead of centralized bureaucratic central planning, what is needed is a franchise organizational structure. In this manner local fire managers are wed to manageable parcels of land to make the everyday decisions, while under the supervision of clear, defined rules of competent land management practice.

CHAPTER ONE

FIRE IN GEOLOGIC HISTORY

In order to gain perspective on fire's role in nature throughout geologic history we need to be able to understand geologic history from the formation of the earth billions of years ago to the present. Scientists believe that earth formed about 4.5 billion years ago and life appeared on its surface within one billion years. The similarities between today's organisms indicate a single common ancestor from which all species diverged.

Fires began in what is known as the Paleozoic Era spanning roughly from 541 to 252.2 million years and is subdivided into six geologic periods, the Cambrian (541.0 – 489.5 million years), Ordovician (489.5 – 445.2 million years), Silurian (445.2 – 423.0 million years), Devonian (445.2 – 372.2 million years), Carboniferous (372.2 – 303.7 million years), and the Permian (303.7 – 254.2 million years).

After the Paleozoic, is the Mesozoic era (252.2 – 66.0 million years), which is subdivided into the Triassic (252 – 208.5 million years), the Jurassic (208.5 – 152.1 million years), and the Cretaceous (152.1 - 72.1 million years).

Finally we have the Cenozoic era (72.1 million – 11.700 thousand years) and this era is further subdivided into the Paleogene (72.1 – 28.1 million years), Neogene (28.1 – 3.60 million years), and the Quaternary (3.60 million years to 11.700 thousand years). All this is further subdivided, but this should be sufficient to describe fire's major role in geologic history. Keep in mind that these dates vary by up to a few million years depending how different researchers calculate the dates. [9]

Around 1.2 billion years ago alga scum began to colonize the land and by 450 million years the first land plants emerged and became well established. Land plants were so successful that it is believed they contributed to the late Devonian extinction event. Vegetative fires began with the establishment of land based flora in the Middle Ordovician period 470 million years ago that caused oxygen levels to rise to 13% permitting the possibility of wildfire. Wildfire is first recorded in the fossil record in the Late Silurian, 420 million years ago, as charcoalified plants and wildfire ash in the geologic strata.

Except for a controversial gap in the Late Devonian, charcoal has been present in the fossil record ever since. More detailed information on fire in the Paleozoic along with oxygen levels and fossil charcoal can be found in the scientific paper called The diversification of Paleozoic fire systems and fluctuations in atmospheric oxygen concentration. [10]

Some scientists speculate that wildfire created a positive feedback loop contributing to warmer dryer climates more conducive to fire. Fossilized charcoal is found in chunks where plants are preserved in great detail or preserved in soot deposits in lakebeds and river deltas. It is believed that fires in low, scrubby wetlands of the Silurian must have been limited in scope. It was in the Middle Devonian that fire really became widespread according to the fossil record. The Devonian period begins at the end of the Silurian 419.2 million years to the beginning of the Carboniferous period about 358.9 million years ago.

The Carboniferous Period extends from 359.2 million years to the beginning of the Permian Period about 299 million years. It seemed that in the Carboniferous fire became very pronounced because of the high-oxygen, high-biomass of the period. It was at this time that both plants and animals became very adapted to a high fire environment that has continued through the Permian and to this day.

Oxygen levels have been slowly declining since the Carboniferous, but fire has continued to be frequent and widespread all over the world. In some areas native peoples living on the land have increased the frequency of fire for hundreds of thousands of years to improve their livelihood, perhaps countering effects of declining oxygen levels. In Africa and Europe man's influence goes back hundreds of thousands of years. In Australia man's influence seems to have been about 60 thousand years ago, and in the United States plants and animals have adapted to more frequent man caused fire, for tens of thousands of years.

It has not been until modern man, that man's activities began to suppress fire. Any misguided attempt to take man out of the fire equation and revert to less frequent fire before man would completely devastate much of these man enhanced frequent fire ecosystems, perhaps as much or more than fire suppression has already.

Let us begin our study of fire in geologic history with the Paleozoic era. I am going to skip the Cambrian period, the earliest period of the Paleozoic, because if there was any fire at all on land, it did not amount to much. I supposed volcanic fire or lava could have been affecting the microbial mats growing around shorelines in wetlands and uplands causing adaptive changes. If lightning ignited anything, it must not have traveled very far burning as a low oxygen smoldering fire in a dried out microbial mat.

Fire in the Ordovician
(488.3 – 443.7 million years)

The Ordovician Period lasted 45 million years. The area north of the tropics was mostly ocean with most of the earth's land mass collected in the supercontinent Gondwana below the tropics. In the Ordovician most of the world's land mass –

Southern Europe, Africa, South America, Antarctica and Australia formed the supercontinent Gondwana. During the early Ordovician, North America straddled the equator and almost the entire continent was underwater.

In the Middle Ordovician, North America rose above water and a tectonic highland corresponding to the later Appalachian Mountains formed on the eastern seaboard. Western and Central Europe were separated, but were moving north to what is today North America. During the Ordovician, Gondwana shifted toward the South Pole with much of it being submerged underwater. Tetrahedral spores similar to primitive land plants of today have been found from the Ordovician showing that plants had colonized the land at this time. Ordovician image [11] Landmass image [11]

As far as I know there is no fossil record of fire in the Ordovician. We can still speculate on the frequency and intensity of fire during this period based upon observing these same vegetative types of plants and their relationship with fire today. It is believed that plants evolved from green algae into primitive liverworts and mosses similar to those today during the Ordovician. During the Ordovician, the microbial mats, fungus and lichens are being supplemented by early mosses and liverworts which we know today build up dead organic material in wet areas. [12]

During times of drought, these peat bogs dry out as in today's swamps periodically and are set on fire by lightning. The burning out of peat bogs in the Okefenokee Swamp and Everglades, for instance, create open water habitats for plants and animals. This provides a sequence of succession stages as the wetlands once again fill up with dead debris that sinks to the bottom. In this manner nature has used fire to create great diversity of both plant and animal species that occupy these succession stages.

Fire in the Silurian
(443.7 – 416.0 million years)

The Silurian was a time of the melting of large glacial formations. This contributed to a substantial rise in ocean levels and a general stabilization of earth's climate that had previously been subjected to erratic fluctuations. Coral reefs made their appearance in the shallow seas created by the flooding of large parts of the continents creating limestone rock over time. The fossil record shows that vascular plants made their appearance along with the ancestors of spiders and centipedes of today. Silurian image [13] Landmass image [14]

It would appear that bog fires expanded out of the swamps and onto dry land as vascular plants began to build up accumulations of debris on high ground. Also during the Silurian oxygen levels continued to rise to levels above of what we have today so we know conditions were now ripe for more frequent fire. [15]

Plants played a significant role in creating these higher oxygen levels that helped insects and other animals as well as provide needed habitat. It also allowed oxygen levels to get high enough to support fire and flames beyond the smoldering peat fires suppressed by lower oxygen levels.

Wildfire is first recorded in the late Silurian in the fossil record 420 million years ago according to Wikipedia. [16] Fire began to be a natural force to be reckoned with, causing regular vegetative fires in the Silurian. So what were these fires like? It does not look like these fires amounted to much, simply burning out bogs during dry periods, and then extending to higher ground where dead vegetation had built up around the newly evolving land plants.

The ancestors of ferns and horsetails evolved in the Silurian and I have noticed that when the leaves of modern ferns die above ground in wetlands and uplands they burn well. In the spring in the Southeastern United States after a winter fire, the

34

ferns quickly sprout back from beneath the ground with fresh new fronds. Not only does the fire remove the mulching caused by the old dead fronds, it provides new useable nutrients to the sprouting plant helping the plant grow stronger and better.

Fire in the Devonian
(419.2 – 358.9 million years)

The Devonian sees fire and fire adapted plants becoming widespread around the globe. At that time, there was the supercontinent of Gondwana in the south and the continent of Siberia in the north, along with the small continent of Euramerica. [17] Wikipedia states:

> "The Devonian period experienced the first significant adaptive radiation of terrestrial life. Since large vertebrate terrestrial herbivores had not yet appeared, free-sporing vascular plants began to spread across dry land, forming extensive forests which covered the continents." Devonian image [18] Landmass image [19]
> "By the middle of the Devonian, several groups of plants had evolved leaves and true roots, and by the end of the period the first seed-bearing plants appeared. Various terrestrial arthropods also became well-established. Fish reached substantial diversity during this time, leading the Devonian to often be dubbed the "Age of Fish"."

During the Middle Devonian, plants began to grow taller and larger. It was during this period that the first trees evolved. [20] It was also at this time that small trees arose growing to about 6 feet high.   A tree trunk from a tree called Eospermatopteris with an attached crown, was discovered in the state of New York. It was 8 meters tall. In the Late Devonian archaeopteris is the best well known tree and is possibly an ancestor of the conifers. This tree reached a height of 18 meters. By the end of the Devonian, the first seed plants had emerged.

In this article titled Newfound Fossils Reveal Secrets of World's Oldest Forest [21] suggests that trees were already dropping their limbs to the ground providing habitat for animals. The Gilboa Forest dates to 385 million years and these forests were widespread around the world at this time. The article states, "And the trees, like modern palms, likely dropped their branches on a yearly or seasonal cycle, filling the forest floor with woody litter suitable for arthropods such as spiders and insects."

The Cabbage Palm in the southeastern United States, like many other palms around the world having evolved from these early forests, are a very fire adapted species. The trunks are so fire tolerant that they can be almost burnt down by hot periodic fires and still survive. These palms have very flammable fronds that burn very hot, adapted to most likely burn out the competition from other evolving tree species. I suggest the reason these forests were so widespread is because of this very successful adaptive fire trait.

The above article also states, "Stein noted that these early trees played a major role in establishing Earth's first terrestrial ecosystems. ""Trees really dominate those kinds of environments they're found in. They really are the entire fabric in which an entire ecology fits in a terrestrial realm,"'" he said.

Fire in the Carboniferous
(359.2 – 299.0 million years)

The Carboniferous began with vast forests already covering the continents and oxygen levels rose above levels not seen before or since. During the first part of the Carboniferous the coal beds were laid down in warm tropical and subtropical forests that provide much of our energy we use today. The arthropods, the ancestors of modern insects, became very prominent and grew to great size because the higher oxygen levels. This was a time that the amphibians flourished feeding on insects, and sometimes the Carboniferous is called the age of amphibians and early reptiles. [22]

A minor extinction event occurred in the middle of the Carboniferous caused by a change in climate as continents collided becoming the supercontinent Pangaea. This caused cooling glaciation, and low sea level. I suspect this coming together of the continents also put species of plants and animals isolated on different continents into conflict with each other causing extinctions of less adaptable species.

This cooling and drying of the planet caused the tropical rainforest to collapse into fragments and the more arid lands saw frequent wildfires both of high intensity and low intensity partly created by high oxygen levels. Some scientists speculate that wildfires were of very high intensity because of the high oxygen levels. I and others suspect that the higher oxygen levels just led to more frequent less intense fires, as leaf litter caught fire quicker than it normally would. This was a time of great adaptation to fire by most species of plants and animals because fire was so frequent and intense, perhaps even more than anytime in earth's history. Carboniferous fire image [23] Landmass image [24]

Fire in the Permian
(303.7 – 254.2 million years)

In the Permian the earth was dominated by the single supercontinent called Pangaea and this supercontinent was surrounded by a global ocean called Panthalassa. According to this Wikipedia entry:

"The extensive rainforests of the Carboniferous had disappeared, leaving behind vast regions of arid desert within the continental interior. Reptiles, who could better cope with these drier conditions, rose to dominance in lieu of their amphibian ancestors. The Permian Period (along with the Paleozoic Era) ended with the largest mass extinction in

Earth's history in which 90% of marine species and 70% of terrestrial species died out. It would take well into the Triassic for life to recover from this catastrophe." [25]

This extinction was so severe that it is the only known extinction where insects were affected. The cause of this mass extinction is still being debated by scientists. Permian image [26] Landmass image [27]

During the Permian many conifer groups, the ancestors or modern day families, spread across Pangaea. Very complex forests were present across Pangaea with a diversity of plant group species. The southern part of the continent saw extensive seed forest ferns and the Ginkgos and cycads evolved during this period. Near the end of the Permian, the archosaurs evolved a group that would give rise to the dinosaurs of the following period.

It is significant that during this period the appearance of the first large herbivores and carnivores came into existence. The reason this is significant (as far as fire is concerned) is that like today, large herbivores are common in fire ecosystems where frequent less intense fires and even occasional catastrophic fires allow for vegetation to grow close to the ground within reach. Otherwise large trees will shade out most of the undergrowth needed to sustain these large herbivores.

Today we know that large herbivores assist fire to expand non-forest fire habitats such as savannas and grasslands by grazing down woody vegetation before it gets so tall that it shades out undergrowth as high brush and trees. The grasses and grasslands had not yet evolved, but we can expect that this environmental niche was already being filled by precursors to grass, as it was by the reptilian herbivore precursors to the mammalian herbivores. In fact, the early ancestors to mammals (the synapsida) were already evolving in the Permian.

Here we come to the end of a brief overview of fire in the Paleozoic and begin our overview of fire in the Mesozoic era starting with the Triassic period.

Fire in the Triassic
(252 – 208.5 million years)

The Triassic is the first period in the Mesozoic era. Major extinction events marked the beginning and end of the Triassic. Wikipedia states:

> "The Triassic began in the wake of the Permian-Triassic extinction event, which left the Earth's biosphere impoverished; it would take well into the middle of the period for life to recover its former diversity. Therapsids and archosaurs were the chief terrestrial vertebrates during this time." Triassic image [28] Landmass image [29]

> "A specialized subgroup of archosaurs, dinosaurs, first appeared in the Late Triassic but did not become dominant until the succeeding Jurassic. The first true mammals, themselves a specialized subgroup of Therapsids also evolved during this period, as well as the first flying vertebrates, the pterosaurs, who like the dinosaurs were a specialized subgroup of archosaurs. The vast supercontinent of Pangaea existed until the mid-Triassic, after which it began to gradually rift into two separate landmasses, Laurasia to the north, and Gondwana to the south."

> "The global climate during the Triassic was mostly hot and dry, with deserts spanning much of Pangaea's interior. However, the climate shifted and became more humid as Pangaea began to drift apart. The end of the period was marked by yet another major mass extinction, wiping out many groups and allowing dinosaurs to assume dominance in the Jurassic." [30]

In the Triassic, the seed plants came to dominate the terrestrial flora. In the northern hemisphere of Pangaea, the conifers flourished. In the southern hemisphere, Glossopteris (a seed fern) was the dominant southern hemisphere tree during the Early Triassic period.

At the end of the Permian, oxygen levels plummeted and fire activity subsided. This activity coincided with the mother of all extinctions that produced a rarity of charcoal in the geologic record. This rarity of charcoal suggests that there was very low biomass throughout the Triassic.

With oxygen levels and biomass both low, we can expect that there was reduced fire activity during this period. However, it was not enough to stop the process of fire adaptation among plants and animals that had already become very extensive by this time. Fires would appear to be limited and contained until oxygen levels rose, with fire increasing exponentially in the late Jurassic through the Cretaceous, as evidenced by huge increases in fossil charcoal.

Fire in the Jurassic
(208.5 – 152.1 million years)

As the Jurassic period began, the supercontinent Pangaea was rifting into the two landmasses, Laurasia to the north, and Gondwana to the south. This breaking apart of Pangaea created more coastlines and changed the climate from dry to humid with many arid deserts of the Triassic becoming lush rainforests. During this period, the dinosaurs flourished and dominated other species such as the early mammals. This was also the period when the first birds appeared. The oceans were filled with marine reptiles such as ichthyosaurs and plesiosaurs. Pterosaurs were still the dominant flying vertebrates. Jurassic Image [31] Landmass image [32]

The conifers dominated the flora as they had during the Triassic with most large trees being conifers. The gymnosperms (the ancestors of modern conifers, cypress, and pines) became very diverse in the Jurassic. The extinct Mesozoic conifer family (Cheirolepidiaceae) dominated low latitude vegetation as well as the Bennettitale shrubs. Cycads, Ginkgos and tree ferns were also common in these Jurassic forests. Ginkgos were common in the middle and higher

40

latitudes in the Northern Hemisphere but rare in the Southern Hemisphere. [33]

Gymnosperms of the past like modern day pines, conifers and cypress, are very fire adapted and are a product of high fire environments. Gymnosperms have resin that is an antimicrobial material that seals wounds from insects, but it is also very flammable. The following link on gymnosperms states:

> "It is also flammable and thus it turns over the nutrients faster and clears the underbrush so that it makes sure to have enough water available for survival by eliminating the competition. The gymnosperms do not burn due to their often 1 foot thick cork that is fire resistant and helps to insulate the phloem against freezing in the winter." [34]

Some gymnosperms like today's redwoods release seeds after cool periodic fires, while some pines have cones that only release seeds after a catastrophic fire. The cones of these pines explode like popcorn when heated and so disperse their seeds on the fire cleared ground. Here the young seedlings can take root because the competition has been burnt away. They also utilize the ash as fertilizer to grow fast and get a jump on the competition that will be sprouting back with their own fire strategies.

Fire in the Cretaceous
(152.1 - 72.1 million years)

The Cretaceous spanned about 79 million years and generally had a warm climate making for high sea levels and many numerous shallow seas. At this time new groups of mammals and birds appeared along with flowering plants. The flowering plants (angiosperms) according to DNA evidence had existed since the Permian, but it was during the Cretaceous that the angiosperms really began to dominate the gymnosperms. Most of these two groups were already very fire adapted from earlier times. The Cretaceous ended with a massive extinction that resulted in the loss of all the non-avian dinosaurs and the large marine reptiles. Cretaceous image [35] Landmass image [36]

It was during the Cretaceous that the breakup of Pangaea became almost complete, with only the continent of Australia still connected to Antarctica. Already Pangaea had broken into Laurasia in the north and Gondwana in the south, with North America pulling away from Eurasia in the Jurassic to become completely separate in the Cretaceous. South America split off from Africa, from which India, Australia and Antarctica were also separating, leaving India adrift in the Indian Ocean. When the Cretaceous ended, most of our present day continents were separated by large oceans such as the North and South Atlantic Ocean. [37]

This breakup of Pangaea was a very important development in the history of life and of fire. Plants and animals had competed together throughout the whole of Pangaea, but when the continent broke apart all these very advanced fire ecosystems became separated. They continued to evolve under more isolated conditions, but the basic structure of plant and animal species had pretty much become set. If one travels about the world one cannot but be impressed by how similar the trees, palms, pines and conifers all look as well as other plant and animal species.

We see very similar fire adaptations throughout all these now separated fire adapted ecosystems. In a sense, man over the past several hundred years has brought the continent of Pangaea back together again by transporting plants and animals back and forth across continents and thus putting separated ecosystems back into contact and competition with each other. It's almost as if man himself became a land bridge between continents for better or for worse.

Today, the more adaptable species moving between continents are in competition with the more specialized less adaptable species. This is resulting in mass extinctions of plant and animal species that lose out in the process. In addition, man is altering the global climate helping more adaptable species, while at the same time destroying more specialized less adaptable species.

It's becoming clear that all these mass and minor extinctions over geologic time mean that nature is intent on limiting or balancing excesses of over specialization and or generalization. This continuous resetting of this biological clock is pitting diversity against adaptability making for optimal survivability regardless of changing environmental conditions.

In the Cretaceous, even the polar regions were free of continental ice sheets, and the land was covered by forests and savannas. Dinosaurs were common in Antarctica even with its long winter night. Of greatest importance was that angiosperms (flowering plants) developed and thrived during the Cretaceous and today represent 80% of the known green plants now living. Encyclopedia Britannica states:

> "The variety of forms found among angiosperms is greater than that of any other plant group. The size alone is quite remarkable, from the smallest individual flowering plant, probably the watermeal (Wolffia; Araceae) at less than 2 millimeters (0.08 inch), to one of the tallest angiosperms, Australia's mountain ash tree (Eucalyptus regnans; Myrtaceae) at about 100 meters (330 feet)." [38]

Flowering plants in all their diversity form the bulk of the vegetation upon which animals feed. This includes the grasses (graminoids) that are usually herbaceous plants with narrow leaves growing from the base. This includes the true grasses as well as sedges and rushes.

Grasses became widespread during the latter part of the Cretaceous and fossilized dinosaur dung has been found with grasses related to modern rice and bamboo. Grasses are the most widespread plant type providing food and energy for all kinds of wildlife and organics. Grasses have adapted to

43

conditions in lush rain forests, dry deserts, mountains and intertidal habitats. [39]

Most grasses have adapted to fire and need fire to exist and flourish. In addition, besides forests and savannas in our global ecosystems, we now have grasslands that have evolved from the fiery Cretaceous. There has always been a close association with large grazing herbivores and their predators with grasslands. Grazing pressures woody vegetation that combined with fire expands grasslands and grassland savannas way beyond what climate conditions and landscape would normally dictate.

During the Cretaceous grasslands evolved along with the grazing dinosaurs and when the dinosaurs were wiped out (except for the birds) grazing mammals evolved to fill the niche left behind by the loss of the dinosaurs. To a limited degree, some birds evolved into this niche of grazing on grasslands like geese and flightless birds.

It seems that a lot more fire research has been done on the Cretaceous than on previous periods. There does seem to be a bias by uninformed researchers as to the role light frequent fires play in maintaining healthy ecosystems. They tend to play up catastrophic fires and downplay light regular periodic fires. I think this is just a carryover of the fire suppression culture of the past century. Catastrophic fire has been a part of fire ecosystems among some plant species usually in cooler climates where it takes longer for fuel accumulations to build up and burn, but over most of the earth light fire burns the ground cover clean before these fuel accumulations can build up to catastrophic proportions.

The scientific paper called Cretaceous wildfires and their impact on the Earth system has this to say about the role of fire in the Cretaceous in its abstract:

"A comprehensive compilation of literature on global Cretaceous charcoal occurrences shows that from the Valanginian on throughout the Cretaceous, terrestrial sedimentary systems frequently preserve charcoal in abundance. This observation indicates that fires were widespread and frequent and that the Cretaceous can be considered a "high-fire" world."

"This increased fire activity has been linked to elevated atmospheric oxygen concentrations, predicted as in excess of 21% throughout this period and 25% during some stages. This extensive wildfire activity would have affected the health, composition and structure of the vegetation and, through habitat loss, probably the fauna. For these reasons, fire activity should be taken into account in Cretaceous vegetation and climate models." [40]

Cenozoic Era
(72.1 million – 11.700 thousand years)

While the end of the Cretaceous period and the beginning of the Cenozoic Era was marked by mass extinctions, it allowed mammals a chance to evolve into the niches that dinosaurs had left. In the plant kingdom regardless of what happened to individual species, the gymnosperms (conifers, pines and cypress), the angiosperms (flowering plants) continued to thrive and flourish evolving to where they are today. It amazes me that there exists today so many types of plants that are changed little from their ancestors from that time plants first colonized the land. Cenozoic image [41] Landmass image [42]

In the Paleogene (the first period of the Cenozoic Era), large grassy plains came into existence in the Eocene (an early division of the Paleogene). The Evolution of Plants article has this to say:

"Grass has the special quality of being more resistant to grazing than other angiosperms. Ancient horses lived in the jungle, but later on new horse species began to set out for more open spaces and they started to eat grass. Probably the large grassy plains developed by co-evolution of grazing animals and grass. This means that there was in interaction between the grazing animals and the grass: the animals ate the grass, the grass became adapted for the sake of survival, the animals adapted in their turn to innovation of the grass, and so on." [43]

As I have stated previously, fire was another prime factor in grassland development as it has been previously in both forest and savannas. The author of the Evolution of Plants article also made the point that to fully understand angiosperms and I assume gymnosperms as well, that an extensive knowledge of the now living flora is necessary because the fossils can be compared to their living relatives. According to this author, the average lifespan of a species is only about 6 million years. I guess where one defines the end of one species and another begins is a bit subjective.

The species from a few million years ago are therefore very similar to the plants we have today and so for the ecosystems as well. In this last part of the Cenozoic Era, the ancestors of man come into the fire and ecosystem equation. We shall continue the discussion of this in the next chapter (Fire and Man) beginning with the hominids several million years ago. Late Cenozoic Image [44]

CHAPTER TWO

FIRE AND MAN

Cenozoic Era
(72.1 million – 11.700 thousand years)

Throughout the Cenozoic all the continents continued to separate. In the Cenozoic the last holdout, Australia, broke free of Antarctica moving north to where it is today. This caused major climate shifts for both Australia and Antarctica as had already happened to other continents after the breakup of the Pangaea. Australia began to dry out as it moved north toward the equator becoming arid in the interior with forest plant species still able to hang on around the periphery of the continent. Antarctica was thrown into a deep freeze because cold ocean currents could now circulate around the whole continent without being blocked by Australia.

With the evolution of fire grassland ecosystems in the early part of the Cenozoic Era and the extinction of the non-avian dinosaurs, the stage was set for the evolution of mammals. The early mammals were now able to expand and diversify to fill the environment niches left behind by the dinosaurs. This diversification led to the evolution of primates, hominids, other species of Homo, and finally, Homo sapiens. If the asteroid had not hit and knocked the dinosaurs and reptiles back, it's quite possible that an intelligent two legged dinosaur would have evolved to fill the environmental niches opening up on the land. At the end of the Cretaceous there were already some small dinosaur raptors that were already growing rapidly in intelligence.

One cannot but be impressed about the way nature's ecosystems and geology evolve new species. On the one hand we have this steady movement forward toward niche specialization in ecosystems, but on the other hand nature seems to roll the dice to create indeterminate outcomes. As our understanding of how evolution works on worlds other than our own increases, I think we will be just amazed by the diversity being created in the universe. I suspect there are worlds where other intelligent predators develop the technology for space travel and leave their home worlds behind, just as we are now doing.

Homo sapiens evolutionary linage can be traced back to the Late Cretaceous period. According to genetic studies primates diverged from other mammals about 85 million years ago, but the earliest fossils appear in the Paleocene around 55 million years ago. According to Wikipedia, the family Hominidae diverged from the Hylobatidae (gibbon) family 15 -20 million years ago. Around 14 million years ago, the orangutans diverged from the Hominidae family.

Bipedalism is the basic adaption of the hominin line. The earliest bipedal hominin is considered to be either Sahelanthropus or Orrorin, with Ardipithecus, a full bipedal, coming later. The gorilla and chimpanzee diverged around this same time, about 4-6 million years ago with either Sahelantropus or Orrorin our last shared ancestor. The early bipedals evolved into the Australopithecines, and later the genus Homo. Ardipithecus image [45] Wikipedia states:

"The earliest documented members of the genus Homo are Homo habilis which evolved around 2.3 million years ago, the earliest species for which there is positive evidence of use of stone tools. The brains of these early hominins were about the size of that of a chimpanzee. During the next million years a process of encephalization began, and with the arrival of

Homo erectus in the fossil record, cranial capacity had doubled to 850 cm." This puts the genus Homo squarely in the Paleolithic Era (2.6 million to 12,000 years) of geologic history. Homo habilis image [46]

"Homo erectus and Homo ergaster were the first of the hominina to leave Africa, and these species spread through Africa, Asia and Europe 1.3 to 1.8 million years ago. It is believed that these species were the first to use fire and complex tools. According to the Recent African Ancestry theory, modern humans evolved in Africa possibly from Homo heidelbergensis, Homo rhodesiensis or Homo antecessor, who migrated out of the continent some 50,000 to 100,000 years ago, replacing local populations of Homo erectus, Homo denisova, Homo floresiensis and Homo neanderthalensis."

"Archaic Homo sapiens, the forerunner of anatomically modern humans, evolved between 400,000 and 250,000 years ago. Recent DNA evidence suggests that several haplotypes of Neanderthal origin are present among all non-African populations and Neanderthals and other hominids, such as Denisova hominin may have contributed up to 6% of their genome to present day humans."

"Anatomically modern humans evolved from archaic Homo sapiens in the Middle Paleolithic, about 200,000 years ago. The transition to behavioral modernity with the development of symbolic culture, language and specialized lithic technology happened about 50,000 years ago according to many anthropologists although some suggest a gradual change in behavior over a longer time span." [47]

Of particular interest to us is that Homo erectus and Homo ergaster were the first species to use fire and complex tools. What this means is that prior to this other earlier species may have been reacting and adapting to fire environments, but not actively using fire to manipulate those environments. This in spite of the fact some ancestors were already making and using stone tools before the time of Homo erectus and Homo ergaster. [48]

I would not be surprised that evidence will continue to surface pushing back the use of fire by various species of Homo to manipulate their fire environments. If these earlier species were manufacturing stone tools, I suspect they may have been using fire to their advantage. I think it's possible that Homo erectus and Homo ergaster were using fire 1.3 to 1.8 million years ago and maybe even Homo habilis as early as 2.3 million years ago.

Fire is much easier to handle and manipulate than people realize. If one has the intelligence to make stone tools and weapons, it's easy enough to take a little grass and spread a light backfire around to better collect insects, or light a hot head-fire with a ember to move game grazing in dry grass to where they can be easily ambushed.

Under my father's watchful eye when I was five years of age, I was already playing around with fire in this manner when we were out burning the woods and fields of our Birdsong Plantation together. Chimpanzees and other animals have surprised researchers in the past with their tool making abilities, causing researchers to rethink what it means to be human. So I don't think it is too farfetched to speculate that our early ancestors might have been using fire in simple ways much further back in time.

Regardless, if Homo erectus was using fire for warmth, protection from predators and cooking, he-she was also using it to help forage and hunt game. This would be sure to increase the low intensity fires in these already very fire prone environments. This in turn would lead to increased light fire and catastrophic fire adaptations and expansion of grasslands and savannas into forested areas. It is suspected that this might well have affected the climate in Africa at the time.

Modern man is no different. He is still pushing back the forest ecosystems in favor of grassland and or agricultural ecosystems and is still using fire as one of the tools to do this as in Africa and South America. This has already been done in North America and Europe for thousands of years as well and is called slash and burn agriculture. I saw this in practice when I was in Southern Mexico as a boy. Trees and brush were cut down and burned, and crops planted right away in the ash fertilized soil. When the soil wore out the indigenous people moved to a new spot. This agricultural practice created a diverse mosaic of successionary stages of plants and animals in areas that would not be normally be prone to fire.

My father in his paper, The Use of Fire: An Historical Background, published in the first Tall Timbers proceedings in 1962 devotes the first paragraphs of his paper to point out that Man is a creature mostly of grasslands. Ed Komarek Sr. states:

> Fire in nature and its use by man has been the subject of much controversy throughout the world for many years. I, however, cannot help but be impressed, after a perusal of world literature on fire, by the great number of excellent studies that have verified the "folk-wisdom" of pioneer peoples and primitive tribes who had to take their sustenance directly from the land. These peoples, generation upon generation, developed knowledge of the use of fire that was akin to "art." This art of the use of fire was not only used for very definite purposes that were valuable to them but was virtually necessary for their existence.
>
> Some historical geographers have pointed out that "Primeval forest is the enemy and not the friend of man; primitive man may make expeditions into the forest but will not settle permanently there" (Hoops; fide East, 1920). Others have used the term "negative" for the influence that forests had to the early spread and occupation of land by man and the term "positive" for open or grasslands. This attitude seems apparent even today for large areas where farm lands have reverted to forest the human population has decreased considerably.
>
> Man throughout his long history has had little use for forest except for fuel and wood for shelter. On continent after continent he fought the forest with fire and other means to increase grassland and field land for pasture and

farming. It was only in the early nineteenth century along with man's early technological development that forest products began to loom large in his economy, and it was only then that he began to protect and replace the forests he had formerly tried so hard to destroy. This awareness of the future importance to him of forests first became apparent in central Europe, and then it spread to northeastern United States and to southeastern India. Curiously enough, these are the only large areas of hardwood deciduous species (beech and maple) in the world where fire can be said to be the most damaging and perhaps of the least possible use in forest management."

In the central United States, both fire suppression and agriculture have led to the loss of the vast fire adapted prairie grasslands. Now forests can grow where there was only grass in the past. We all know the effect man is having on the environment today. However, we are slow to appreciate the effect early man has had on the environment, especially when fire has been used as a tool. Early man used fire for hundreds of thousands of years to reshape the fire ecosystems in which he has lived in order to improve his livelihood just as he does today.

As we can see from the following Wikipedia entry, there is still debate among scientists as to just when man's ancestors began to use fire. I think as more evidence accumulates, it will be proved that man's ancestors used fire even earlier than the one million years indicated here.

Wikipedia states:

"Sites in Europe and Asia seem to indicate controlled use of fire by *H. erectus*, dating back at least 1 million years.[45] A presentation at the Paleoanthropology Society annual meeting in Montreal, Quebec in March 2004 stated that there is evidence for controlled fires in excavations in northern Israel from about 690,000 to 790,000 years ago. A site called Terra Amata, located on the French Riviera, which lies on an ancient beach, seems to have been occupied by *H. erectus*; it contains evidence of controlled fire, dated at around 300,000 years ago.[46]"

"Excavations dating from approximately 790,000 years ago in Israel suggest that H. erectus not only controlled fire but could light fires.[46] Finally, evidence from a site in the Northern Cape Province of South Africa is consistent with controlled fire use by H. erectus one million years ago.[45] Despite these examples, some scholars continue to assert that the controlled use of fire was not typical of H. erectus, but only of later species of Homo, such as H. heidelbergensis, H. neanderthalensis, and H. sapiens)." [49]

Another Wikipedia entry called Control of fire by early humans is more conservative, but still indicates debate on evidence for controlled fire going back 1.8 million years by man's ancestors. I think we can safely say that fire use goes back hundreds of thousands of years if not millions.

As this Wikipedia entry points out, the control and use of fire as a tool was a turning point in the cultural aspect of human evolution that allowed humans to cook food and obtain warmth and protection. It also allowed the expansion of human activity into the dark and colder hours of the night to provide protection from predators and insects. I would also add that fire as a tool could easily be used to alter and manipulate the high fire ecosystems in which these ancestral humans lived, especially in the fire grasslands and savannas of Africa in which humans first evolved. This more conservative Wikipedia entry says that:

"Evidence for the controlled use of fire by Homo erectus beginning some 400,000 years ago has wide scholarly support, while claims regarding earlier evidence are mostly dismissed as inconclusive and sketchy. Claims for the earliest definitive evidence of control of fire by a member of Homo, range from 0.2 to 1.7 million years ago." [50]

The controversy around when man's ancestors used fire can become quite heated because of the complexity involved in interpreting the evidence. The article (Who Mastered Fire) discusses this controversy in depth. [51]

Some scientists such as Richard Wrangham of Harvard University believe the control of fire and cooking explains the increase in hominid brain sizes, smaller digestive tract, smaller teeth and jaws and decrease in sexual dimorphism that occurred roughly 1.8 million years ago. Richard says that raw meat and vegetables could not have provided the necessary energy to support the normal hunter-gatherer lifestyle. Other anthropologists in the mainstream dispute this claim believing that human brain-size occurred well before the advent of cooking, due to a shift away from the consumption of nuts and berries to the consumption of meat.

The teeth of Homo erectus do show gradual shrinking suggesting a transition from crunchier foods to softer foods such as meat and various cooked foods. The evidence of cooking comes from blackened animal bones found at various archaeological sites.

Recently an international team in 2012 found evidence that Homo erectus used fire for cooking one million years ago. This international team was led by the University of Toronto and Hebrew University and has identified the earliest known evidence of the use of fire by human ancestors according to Science Daily. Microscopic traces of wood ash, alongside animal bones and stone tools, were found in a layer dated to one million years in the Wonderwerk Cave in South Africa.

"Analysis of the sediment by lead authors Francesco Berna and Paul Goldberg (pictured below right) of Boston University revealed ash plant remains and burned bone fragments both of which appear to have been burned locally rather than carried into the cave by wind or water. The researchers also found extensive evidence of surface discoloration that is typical of burning."

""The control of fire would have been a major turning point in human evolution," said Chazan. "The impact of cooking food is well documented, but the impact of control over fire would have touched all elements of human society. Socializing around a camp fire might actually be an essential aspect of what makes us human."" [51]

Searching the internet I find that there have been recent finds that are filling in the evidentiary base for fire use at least a million years as in a recent BBC, Early human fire skills revealed article. Author Paul Rincon says:

"Human-like species migrating out of their African homeland had mastered the use of fire up to 790,000 years ago, the journal Science reports. The evidence from northern Israel, suggests species such as Homo erectus may have been surprising sophisticated in their behavior." [52]

Modern human "blacksmiths" in Africa have been found to have been using fire to manufacture stone tools at least 72,000 years ago according an article in Telegraph.. "We show that early modern humans at 72,000 years ago, and perhaps as early as 164,000 years ago in coastal South Africa, were using carefully controlled hearths in a complex process to heat stone and change its properties, the process known as heat treatment." [53]

So increasingly it looks as if man's ancestors have been using fire over a million years, not just for cooking and protection, but I and others suspect also to alter and shape the fire landscape in which they lived. It might have been as simple as children playing with fire brands letting the fire get away into the surrounding dry grass, or much more complex behavior using fire to herd game and insects.

Even today in some parts of Africa, native peoples burn small patches of dead grass in order to attract game. They then place snares and traps around the area to capture the game when it comes into the burnt patch to feed on the fresh green grass from the fire. We also have modern day accounts where native peoples use the fire to herd animals into areas where they can be captured and killed.

The Cenozoic Era ends 11.700 thousand years ago with modern man firmly entrenched on all continents except Antarctica. By now man has evolved to the point where he no longer reacting and adapting to nature, but he is actively manipulating global ecosystems to his own ends. Fire has become a major tool used by Homo sapiens to create and manipulate fire ecosystems to his advantage both knowingly or unknowingly. During the early stages of agriculture, man learns to genetically manipulate fire grasses for food through the process of selection, and learns to domesticate and herd wild animals for his livelihood as well.

The Holocene
(11,500 years – present)

The Holocene is considered a warm period, an interglacial period in the current ice age. Wikipedia states:

"The Holocene also encompasses within it the growth and impacts of the human species world-wide, including all its written history and overall significant transition toward urban living in the present." [54]

According to the scientific paper, Holocene biomass burning and global dynamics of the carbon cycle, "Fire regimes have changed significantly during the Holocene due to changes in climate, vegetation and in human practices." "In Europe the significant increase of fire activity is dated to approximately 6000 years ago. In north-eastern North America burning activity was greatest before 7500 years ago, very low between 7500 -3000 years, and has been increasing since 3000 years ago."

"In tropical America, the pattern is more complex and apparently latitudinally zonal. Maximum burning occurred in the southern Amazon basin and in Central America during the middle Holocene and during the last 2000 years in the northern Amazon basin. In Oceania, biomass burning has

56

decreased since a maximum 5000 years ago. Biomass burning has broadly increased in the Northern and Southern hemispheres throughout the second half of the Holocene associated with changes in climate and human practices."
(link)

As I search the Internet I see a huge amount of specialized information relating to indigenous peoples increased frequent light fire activities during the Holocene as well as a drop off of frequent fire activity in favor of more catastrophic fire activity during the period of European colonization. One such example is the paper, Reconstructing Holocene fire history in a southern Appalachian forest using soil charcoal.

The authors state:

> "Summed probability analysis taking into account radiometric errors, suggests that fires became more frequent approximately 1000 years ago, coinciding with the appearance of Woodland Tradition Native Americans in this region." [55]

It's really good to see scientists obtaining proof as to the decrease of light intensity fire and an increase of high intensity fire during and after European colonization. Mitchell Power writes of an attempt to do just that in 2013. He states:

> "During the Holocene, the last 11,000 years, climate, vegetation, and likely, humans have been key controls to changing fire regimes in the Americas. A long-accepted paradigm is that of the 'noble savage', whereby indigenous peoples lived in harmony within a pristine wilderness, with little or no significant impact upon natural ecosystems. However, increasing evidence for extensive, large-scale landscape modification is leading many archaeologists to argue that the very notion of 'virgin' forests is a myth, and that prior to the Spanish Conquest, forests, grasslands, and savannas were heavily managed using fire, transforming a once pristine wilderness into a 'cultural parkland' (Heckenberger et al. 2003)."

"According to this theory, the 'pristine wilderness' first encountered by Europeans was in fact secondary forest recovering after the catastrophic crash in indigenous populations caused by first exposure to European diseases that swept through the Americas in advance of European settlers (Mann, 2006). If true, then fire frequencies would be expected to be significantly lower in the 16 and 17th centuries compared with the 15th century. We aim to test this hypothesis using data from the recently created Global Charcoal Database, analyzing charcoal data from throughout the Americas." [56]

Jenn Marlon at this same Internet site indicates a broad scientific consensus developing as to the damage being done to fire environments and to mankind by increased fire suppression in the 20th century and into the 21th especially in places like the Western United States. The question is what are we going to do about this sad state of affairs?

Marlon states:

"Since the late 1800s, human activities and the ecological effects of recent high fire activity caused a large, abrupt decline in burning similar to the LIA fire decline. Consequently, there is now a forest "fire deficit" in the western United States attributable to the combined effects of human activities, ecological, and climate changes. Large fires in the late 20th and 21st century fires have begun to address the fire deficit, but it is continuing to grow."

A Wikipedia entry on Historical Ecology continues to drive the point home that man has been a continuing and intrinsic part of global fire ecosystems for a long time. Any careless and foolhardy attempt to return ecosystems to a state before man will lead to do even more damage to fire ecosystems and the plants and animals that have become fire adapted.

"Both destructive and at times constructive, anthropogenic fire is the most immediately visible human-

58

mediated disturbance, and without it, many landscapes would become denatured.[16] Humans have practiced controlled burns of forests globally for thousands of years, shaping landscapes in order to better fit their needs. They burned vegetation and forests to create space for crops, sometimes resulting in higher levels of species diversity."

"Today, in the absence of indigenous populations who once practiced controlled burns (most notably in North America and Australia), naturally ignited wildfires have increased. In addition, there has been destabilization of "ecosystem after ecosystem, and there is good documentation to suggest fire exclusion by Europeans has led to floral and faunal extinctions."[11]

[57] (Historical Ecology)

I am glad to see in this UN study on global fire management called Fire management – global assessment 2006, the recognition that man is an intrinsic part of past and present ecosystems. It also stresses the importance of prescribed fire in managing these ecosystems all over the world. In the forward Peter Holmgren states:

"This study presents information on fire in greater depth than was possible in FRA 2005, including its incidence, impact and management in different regions of the world. It recognizes that not all fires are destructive, and that fire management is an essential part of sustainable forest management. Indeed, some ecosystems require fire to induce regeneration and to maintain or enhance biodiversity, agricultural productivity and the carrying capacity of pastoral systems. The study also finds that people are the overwhelming cause of fires in every region, for a wide range of reasons."

"Much more must be done to help the general public and policy-makers understand the scale of this threat and take long-term preventive action, not simply emergency suppression measures when a fire disaster strikes. More must be done, as well, to improve the understanding of fire by urban people at the wildland/urban interface, especially the need to reduce fire threat through fuel management, including prescribed fire burning."

It's important to reconstruct indigenous people's use of fire and their close relationship to their fire environments. With today's concentration of populations in the cities, people have lost a lot of this connectedness to the land. Sure one can venture out into the wilderness from time to time, but how close are you really getting.

I was raised on a plantation in the Deep South where I lived and played on the land just about every day. I went over this square mile of land and stepped on just about every square yard of it at one time or another. I got to intimately know most every tree, animal, and plant so when I controlled burned, I was very careful to not unnecessarily scar a pine tree or damage even a small ecosystem.

Leon Neel has pointed out in his book The Art of Managing Longleaf, how well some Quail Plantations in the Deep South manage their fire ecosystems right down to single pine trees. This book presents a model that government fire managers should emulate based on a deep spiritual connectedness to natural ecosystems, due to spending so much time on individual properties. This is what native peoples for thousands of years even hundreds of thousands of years have been doing.

How different it is today where government fire managers stream out of the city to burn public lands on an industrial scale, dressed in heavy boots, gloves and helmets and clothing like environmental storm troopers. ☺ These troopers find themselves cast in a losing battle, underfunded and trying to use prescribed fire in a short fire season window. To make matters worse they have to comply with increasingly stifling EPA smoke regulations.

How can these land managers be expected to effectively manage micro fire habitats? How can they be concerned about individual trees or small plant communities? It's as if they are forced to make war on fire with fire, rather than develop a friendly relationship with fire and the fire environment.

In order to understand how native peoples got it right, let's take more of a look at individual countries and regions to understand how native peoples not only coped and adapted to fire, but how they lived on the land and managed fire for a livelihood during the Holocene. This knowledge and wisdom of native peoples must be incorporated into the restructuring of our present dysfunctional, over centralized government land management bureaucracies.

Man and Fire in Africa

The climate in northern Africa at the beginning of the Holocene had a moist climate that supported tropical vegetation and this period continued to about 4300 years ago. At that time North Africa began to dry out creating the Sahara Desert where there was once freshwater lakes, hippos and crocodiles. In Southwestern Africa the climate fluctuated between humid and dryer conditions from the beginning of the Holocene to about 3500 years when the climate became more arid. Overall this drying out of Africa meant shifting fire ecosystems as well, and it became necessary for human African populations to adapt, both in hunting-gathering societies and in agriculture societies.

Africa is considered the fire continent by ecologists because of widespread annual burning by man. Today 90% of all African fires are ignited by man's controlled burns and about 10% by lightning. The most extensive area of fire grasslands and savanna in the world are in the southern, western, and eastern regions of Africa. These regions become extremely flammable in the dry season which is from May to October in Southern Africa, and January to April, in west and east Africa.

Stephen Pyne correctly points out in his World On Fire article that:

"Unlike human-caused burning in regions of the world that are not naturally given to a fire ecology, such as the South American tropical rainforest, fire is an integral component of the African ecosystem that scientists believe has an extensive and important evolutionary history on the continent. Many African plant species and animals, for example, have growth and reproduction cycles so linked to Africa's fire seasons that they would likely become locally extinct without fire." [58]

Subsistence hunting using fire by man and his ancestors has surely been going on in Africa for at least hundreds of thousands of years and little unchanged from techniques used today. The following is from an article, Wildlife and food security in Africa that discusses subsistence hunting and gathering in Africa.

"The use of fire in group hunting is more common in the grassland savannah areas. Members of the group are positioned strategically around a patch of grassland known to contain wild animals. The area is then set on fire and animals are killed with cutlasses and clubs as they run out of the area to escape the fire."

"Within the forest areas, fire is regularly used to smoke out rodents such as the giant rat Cricetomys gambianus from their burrows. A group of rat hunters would search for rat holes and set fire at the entrance using palm branches and dry leaves. The smoke penetrates the burrow and forces the rat to come out. In the meantime, members of the group would be waiting at strategic points around the burrow ready to kill the rat as soon as it comes out. Often the animal dies in the burrow out of suffocation from the smoke, in which case, it is dug out." [59]

There are remnant populations of Bushmen in Africa that still live primarily by hunting and gathering, but who now contract with tourist operators. They are allowed to hunt and forage in the big game parks of Tanzania. The Hadza live in central Tanzania around Lake Eyasi in the central Rift Valley and on Serengeti Plateau. According the article Enduring Cultures:

"The Bushmen people raise no livestock grow no food. And live without calendars or rules. They are living a hunter-gatherer existence that is little changed from 10,000 years ago. It is estimated that the Hadza number just under 1000. Today they are the last functioning hunter-gathers in Africa."

Genetic testing seems to indicate that the Hadza are more closely related to the Pygmies than to any other peoples of Africa. They still retain a click language that may go back 40,000 years. None of their neighbors have this click language which again argues for their isolation from other tribes in the region. The Enduring Cultures article says:

"Hadza men hunt individually or in groups and use simple bows and poisoned arrows to hunt small and large game. If the catch small animals – such as squirrels or birds – they will light a fire and toss the carcass into the flames to cook and eat it then and there."

It's not hard to see that such fires could easily spread to surrounding vegetation if the people were careless, or deliberately being used for hunting of game in these fire environments. [60]

In the South African Kalahari the San people still live where their ancestors were hunter – gathers for thousands of years maybe even tens of thousands of years. The San people are known for their rock art and there are 15000 known rock art sites in their region. This rock art is one of the archaeological wonders of the world and it expresses the religious and spiritual beliefs of the San people.

According to this article on South African history:

"The most important part of the San's lives is fire. The men are responsible for making fire and use two fire sticks that they carry with them at all times. They would rub the two sticks together until it makes a spark to light some dry grass. Their lives revolved around their fire because it provided warmth, light and a way to cook food. The family would also hang their possessions in bushes close to the fire."

The San people developed their own type of bow and arrow and used poisoned heads on the arrows. They stalk a large animal by following the tracks and then shoot the animal with the poison arrow. The narrow shaft of the arrow falls to the ground but the poison tip remains in the animal. It takes hours and even days for the poison to slow down the animal where the hunters can get close enough to finish it off with arrows and spears. We can see that these hunting techniques are widespread across Africa, and intrinsic to primitive African culture. The same can be said as to fire use and adaptation to fire environments in which these people hunt and gather food.

Man and Fire Australia

Much of Australia is composed of fire ecosystems that continued to evolve as Australia separated from Antarctica finalizing the breakup of Pangaea and Gondwana. The drying up of Australia's interior evolved major changes to Australia's ecosystems and I am glad to see fire ecologists like Jim Kohen in his paper, The Impact of Fire: An Historical Perspective, stresses the importance of understanding both geologic history and human history in order to best understand today's fire environments in Australia. Jim states:

> "Climatic change and the drying out of the Australian environment resulted in the decline of rainforests and the dominance of sclerophyllous vegetation. Fire has been a major component of this process, promoting those plants which could cope with fire at the expense of those which could not. Aboriginal people added to the firing frequency, but maintained low population densities until the last 5,000

years, when fire became an integral component of their economic system. They used fire as a tool to promote and maintain the vegetation associations which were most productive. Once traditional Aboriginal burning regimes ceased following European settlement, the standing fuel increased, resulting in sporadic but more intense fires."

"The old Gondwana forests had been dominated by gymnosperms - the conifers, auracaria and podocarps, but just as Gondwana broke up, the angiosperms began to radiate. Because of the large size and climatic diversity which existed within Gondwana, the vegetation associations were varied, and when Australia separated it took with it enough angiosperms to expand and almost fill the entire continent. *Nothofagus*, the Antarctic beech, was one of the early angiosperms which were present on Gondwana when Australia became isolated. Minor families included the Myrtaceae, the grasses, the Xanthorrhoeas, and the chenopods (Pyne, 1991). A few of the genera which were present and which later became important included *Eucalyptus, Banksia, Hakea* and *Melaleuca*. A similar vegetation association characterized parts of Antarctica, South America and New Zealand. Where the rainfall was year round and moderately high, rainforests were maintained, but the minor flora adapted and spread, filling many of the niches in the drier areas."

"This regular firing favored not only fire-tolerant or fire-resistant plants, but also encouraged those animals which were favored by more open country. On this basis, it is clear that Aboriginal burning, in many areas at least, did impact on the "natural" ecosystem, producing a range of vegetation associations which would maximize productivity in terms of the food requirements of the Aborigines. Jones goes so far as to say that "through firing over thousands of years, Aboriginal man has managed to extend his natural habitat zone" (Jones, 1969)."

In Australia by 38,000 years ago the Auracarian Rainforest had mostly disappeared with casuarina beginning to be replaced by eucalypts. According to scientists there was a

65

huge amount of charcoal deposited in lakes and swamps and the cause seems to be that the new evolving vegetation was very fire prone. It is interesting that this was not long after Aboriginal people began to settle in Australia using fire as a tool for subsistence living.

According to Kohen, Rhys Jones in 1969 came up with the idea of "firestick farming" by Aboriginal people. Jones was one of the first to suggest that the Aboriginal people were doing controlled burning using light fire to increase the productivity of the land, replacing mature forests with open woodlands and grasslands. Aboriginal traditional burning increased the diversity of their fire environments by creating a mosaic of fire patterns depending on the intensity of the fire and the time of the year the land was burned.

Because the people lived on the land and derived their livelihood from the land, they had an intimate knowledge not only of fire as a tool, but just about every square foot of the land they occupied. This was an intimate knowledge passed down from generation to generation, quite the opposite for today's fire managers of European descent. Where ever we go in the world today this still pretty much rings true. Kohen provides just one more example of European ignorance of fire ecosystems that continues to this day, continuing to devastate Australia's ecosystems and causing much property damage and loss of life due to catastrophic fire.

"When Europeans first settled in Australia in 1788, they found a landscape dominated by eucalypts. Certainly there were some areas of dense vegetation which contained a greater diversity of trees. However, the dominant feature of the landscape was the ubiquitous gum tree. When expeditions began exploring the countryside around Sydney, they encountered a range of vegetation associations very different to those which we see in the National Parks around Sydney today.

On soils derived from Hawkesbury sandstone, Wianamatta shale, Tertiary alluvial deposits, and igneous intrusions, they found environments which reminded them of the manicured parks of England, with trees well-spaced and a grassy understory. Peter Cunningham (1827) described the country west of Parramatta and Liverpool as "a fine timbered country, perfectly clear of bush, through which you might, generally speaking, drive a gig in all directions, without any impediment in the shape of rocks, scrubs, or close forest".

This confirmed earlier accounts by Governor Phillip, who suggested that the trees were "growing at a distance of some twenty to forty feet from each other, and in general entirely free from brushwood ..." (Phillip, 1789). It is clear that it was primarily Aboriginal burning practices which maintained an open environment dominated by well-spaced trees and grass. Once the Aborigines stopped burning, the underbrush returned where none had previously existed. Benson and Howell (1990: 20) suggest that the growth of *Bursaria spinosa* in the Sydney area in the 1820s may be related to a changed fire regime, the cessation of Aboriginal burning."

We can find numerous accounts of Aboriginal burning and hunting practices. Fire by Australia's indigenous people was made in several ways including fire drills, flints, and pyrites. The traditional desert Aboriginal men cut a wedge out of a soft wood and placed tinder in the wedge, soft grass or kangaroo dung. They then took the edge of their spear thrower or boomerang and using a sawing motion ignited the tinder in the wedge. It was the man's job to start the fire and the woman's job to keep the fire burning.

In the Stanford Report called, Aboriginal hunting and burning increase Australia's desert biodiversity, Stanford researchers give some good detail on a group of Martu, indigenous Australians that inhabit the Western Desert. The Martu hunter-gathers light fires to expose the hiding places of their prey like monitor lizards that can grow up to 6 feet long. According to this Stanford Report and Stanford's anthropologists Douglas and Rebecca Bird, these generations-old day-to-day routines have reshaped Australia's Western Desert habitats. [61]

The Stanford Report correctly states:

"In many cases, humans aren't the wrench in nature's gears but an important piece of the clockwork, he added. And because so much of Australia's Western Desert, from lizards to shrubs, revolves around Martu practices, conservation efforts will succeed only if they incorporate traditional goanna-hunting practices, he said. "We're trying to demonstrate what would happen if you did pull people off the landscape," he said. "What happens when you break all of these co-evolutionary links between people who've lived on the landscape for thousands of years and the diversity of the faunal and floral community?"

This article on the Martu discusses the great importance of the Martu's intimate knowledge and respect for the land. This is something we rarely see with today's global public and private land managers. Managers who are trapped by inexperience and limited resources to struggle to just keep accumulating heavy fuel loads down with controlled fire, with no time or resources to really become intimate with the land they are burning.

"Martu life revolves around hunting and fire, Douglas Bird explained. Martu inherit ritual duties that correspond to certain tracts of desert called "estates." An important part of this inheritance is the knowledge of when and where to light smoldering brush fires. Martu never start blazes without knowing every nook and cranny of a territory and often forgo campfires when traveling through foreign estates, he said.

"You never burn unless you're with someone who has all of that knowledge about that estate," he added. "If your fire were to threaten one of those totemic spots where they keep all their religious paraphernalia associated with these rituals, it's technically punishable by death."

The middle-aged and elderly women who typically hunt for goanna can spot the animal's burrows and tracks better in burn scars than in thick spinifex grass, explained Rebecca Bird, an associate professor of anthropology.

Burning desert in about 55-acre chunks, the hunters make their grounds a patchwork quilt of recently burnt earth and recovering vegetation. These scars are much smaller than those left by lightning wildfires, which char an average of 2,000 acres. Burning back grasses and other fire-prone plants encourages the growth of a diverse range of annual vegetation, she said. The variable turf of Martu hunting grounds allows small mammals to find plenty of places to hide from predators, she added, while areas free of human burning lack this patchwork quality and are home to fewer plants and animals."

One of the good things about Africa is that unlike North America and Australia, European fire exclusion policies really never caught hold. Furthermore, the native fire managers were never devastated as badly as in the rest of the world from disease. Throughout the 20th century the native fire managers in Africa continued to use fire as they always had done. They burned much of Africa with light periodic fire, while forests, savannas and grasslands elsewhere stagnated, decayed and exploded with catastrophic fire because of stupid European fire suppression policies.

Man and Fire in Asia and India

In the Cenozoic, India collided with Asia 55 to 45 million years ago and Arabia collided with Eurasia closing the Tethys ocean about 35 million years ago. The continent of India drove itself into Asia forcing up the Himalayan Mountains and finally South America became attached to North America.

Homo erectus left Africa and populated China as early as 1.66 million years based on stone tools found in the Nihewan Basin. Stone tools found at Xiaochangliang were dated to 1.36

million years. The archaeological site of Xihoudu in Shanxi Province is the earliest recorded use of fire by Homo erectus, which is dated at 1.27 million years. Southeast Asia was reached 1.7 million years ago and then West Europe 1.2 million years ago. [62]

It seems that there were two migrations of Homo sapiens out of Africa that spread out and displaced Homo erectus. One was across the Red Sea traveling along the coastal regions to India and the other from East Africa heading north and crossing into Asia through the Sinai. It now seems that one group of humans entered India 70,000 years ago and displaced Homo heidelbergensis before they entered Europe and displaced the Neanderthals. [63]

The first wave of migrations seems to have happened 90,000 to 130,000 years ago and the first populations reached South Asia 50,000 years ago with East Asia (Korea, Japan) reached 30,000 years ago. [64]

So we see that Homo erectus using fire could have been altering fire landscapes over a million years ago, but just how much effect there was is hard to determine. In Asia as elsewhere the effects of Homo sapiens seems to have been pronounced using fire to expand the fire ecosystems to which he was so adapted.

A cursory investigation of the role of fire in India's fire ecosystems shows that India is far behind in understanding the beneficial role of fire in nature especially among government policymakers. India has strict laws against lighting forest fires seemly inspired by the European colonizers, but native peoples still go about their burning for their livelihood as they have since they arrived on the Indian continent.

I have found some lonely pro-fire articles that remind me what it was like when my father rounded up global pro-fire mavericks in face of powerful US Forest Service government opposition in the 1950s and 1960s. In India as in the Western United States, unknowledgeable land managers and politicians scapegoat fire for man caused problems. In India the real problem is not fire, but overpopulation and overgrazing.

The native peoples of India are just doing what they have always done to make a living off the land and that is pushing back the forest using fire. The problem with the severe loss of forest land in India should not be blamed on fire, no more than man caused clear-cutting, excessive logging, and increasing fuel load accumulations in the Western United States be blamed on fire. The article, A Fiery History pretty much sums up the situation in India. [65]

The authors in this article correctly point out that humans have been using fire since they first arrived on the subcontinent of India.

"Ethnographers have compared the centrality of the role fire has played in the life and culture of adivasis in the Andamans—and possibly adivasis elsewhere in India—with the role fire has played in the culture of the Australian aborigines. Fire was also important in the lives of the early Aryans and their livestock, with Agni, the fire god, occupying a prominent place amongst their deities."

"Fire continues to be a forest management tool for forest-dependent communities across the country today. This is so even though the Indian Forest Act of 1927 regards the setting of fires to be a punishable offence and makes it mandatory for all forest dwellers to assist in the prevention and control of fires. People burn for a variety of reasons, whether it is to encourage fresh fodder for grazing livestock, or to aid in the collection of non-timber forest products such as *mahua and tendu*, or to encourage regeneration of particular species."

"However, because of the prevailing ban-and-punish policy, fires today are often set surreptitiously and can become uncontrolled. There is also evidence that with shrinking forest areas and increased demographic pressures there has been a reduction in fire return intervals over the past century. And increasingly, there has been a breakdown of traditional fire management systems (e.g., in the northeast). All of this further fuels the widespread opinion that all fires are destructive and result in forest degradation. The reality, however, is probably much more nuanced."

The official policy on fire in India has been fire suppression instituted by Europeans in the late 19th century, who deceptively called it "scientific", when in fact such policy had no scientific basis. This official policy faced skepticism right from the beginning from local people and a few European foresters, who understood correctly from long direct experience that fire was an integral part of forest, savanna and grassland ecosystems of India. In spite of this official policy some controlled burning was strictly controlled, but legally allowed, for forest practices. The general public just ignored the laws and kept on burning as they had for thousands of years though more covertly than before.

According to this article attitudes are slowly changing in India and there are attempts underway to move toward an official recognition that fire is important to India's fire ecosystems. They also point out the necessity of engaging local people who are experienced fire managers with an intimate connectivity to the land.

"Our engagement with the Soliga community in the BR Hills has shown that their knowledge of the ecological role of fires is very sophisticated, and that fire has played an important role in their management of the forest. Far from regarding fire as a destructive force, they recognize a number of benefits of fire for the dynamics of the forests. They also recognize the importance of the timing of fire occurrence, distinguishing beneficial early-dry-season fires—for which they have a particular name, the *tharagu benki*—from the more severe and destructive fires that occur later in the dry season. Given this background, any attempt to manage this landscape, and possibly other landscapes like it, cannot but engage with its fire history."

Gaurav Moghe tells us a little about the Soliga native peoples of India in this entry called, The Soliga community of Karnataka, and their intimate relationship with nature. These people's ancestral lands have now been deemed a wildlife sanctuary. This combined with the general over population problems all over India is putting these native peoples under tremendous stress. [66]

"The Soligas are nomadic people who have lived in the Biligiriranga Hills region of Southern Karnataka for centuries. Soligas - whose name means *Children of the Bamboo* - live off forest produce like honey, berries and timber. They do cultivate a little bit of pigeon peas, beans, pumpkins and millet but mostly for their own consumption [1]. Many Soligas, even today, live in small shelters called *pudus* deep inside the dense forests of Southern Western Ghats. The community is so dependent on biodiversity that they revere Mother Nature - not just the forests, the animals and the trees, but even the land they walk upon [2].

The knowledge that Soligas possess about the forests of the Western Ghats can be used for biodiversity conservation. For example, a study performed in 2008 looked at the Soliga claim that forest fires are in-fact beneficial for biodiversity [3]. Generally, forest fires would be extinguished by the forest department using modern fire-suppression regimes. However, the Soligas claimed that natural fires are inherent part of the forest biome and extinguishing such fires leads to increased parasitic load. The above study looked at infections of *Loranthus* - a plant parasite - on *Phyllanthus emblica (Aamla)* trees and found that fire indeed reduced the parasitic load on these trees and increased their survival [3]. This example points to the importance of considering local, folk knowledge in any biodiversity conservation regime."

China

China is a vast country and is home to one of the world's oldest civilizations and can roughly be divided into North and South China. Since the beginning of the Cenozoic, North China has been impacted greatly by intercontinental rifting and extensional tectonics as the Continent of India drives itself into Asia creating the Himalayan Mountain Range. North China is chilly and composed of flat plains, grasslands and desert. South China is warm and rainy and is composed of lush mountains and rivers. Consequently there is a great diversity of fire ecosystems across China.

China is the home of 1.3 billion people according to the 2000 Census and encompasses a huge variety of climates and landscapes. Its borders extend through Central and South-East Asia, from the Pacific Ocean and tropical jungles to the Himalayas. It contains mostly plateaus and mountains in the east, grasslands and plains in the North and the south is dominated by hills and low mountain ranges.

Much of the population lives in the east along or near the shores of the Yellow and East China Sea and on the plains of the rivers flowing into the Pacific, such as the Yangtze and the Huang He. Because of its size and diversity of geographic features, it has diversity of human cultures and societies throughout antiquity. China has over 100 distinct ethnic groups.

In the lower Yangtze it is believed that rice was first domesticated. The adoption of cereal cultivation out of fire adapted grasses was one of the most important cultural processes in human history and marked the transition from hunting and gathering by Mesolithic foragers to the food producing economy of Neolithic farmers. The article Fire and flood management of coastal swamp enabled first rice paddy cultivation in east China states:

"Here we report detailed evidence from Kuahuqiao that reveals the precise cultural and environmental context of rice cultivation at this earliest known Neolithic site in eastern China, 7,700 calibrated years before present (cal. yr. bp). Pollen, algal, fungal spore and micro-charcoal data from sediments demonstrate that these Neolithic communities selected lowland swamps for their rice cultivation and settlement, using fire to clear alder-dominated wetland scrub and prepare the site for occupation, then to maintain wet grassland vegetation of paddy type."

While Homo erectus made it to China over a million years ago and used fire, the modern human presence has been dated to 67,000 years ago in Guangxi. So we can expect that these Chinese ecosystems have been substantially modified by indigenous peoples using fire for tens of thousands of years for hunting-gathering and for agriculture. Unfortunately, like so much of the rest of the world, European anti-fire policies have spread to China suppressing light fire burning in the 20th century.

The result has been a predictable rise in catastrophic fires that have killed hundreds of people and burnt forests to the ground as in the Western United States and Australia. In the past few decades because of rapid economic development, China now has the resources and capability to suppress natural light fires so essential to most fire ecosystems with the predictable catastrophic fire results. In India and Africa the central governments have not been able to exercise the kind of control over the people necessary to suppress fires and build fuel loads to catastrophic levels.

Here are some examples. The 1987 Daxinganling wildfire also known as the May 6 fire began in northeast Daxinganling Prefecture Heilongjiang on May 6, 1987. It burned almost a month and was stopped on June 2, 1987. The fire covered 2,500,000 acres with about 266 people wounded, 211 dead, and 50,000 left homeless. The 2010 Dawu fire was a grassland fire that killed 22 people. It was short and lasted only 17 hours and 30 minutes with the fire occurring in extreme terrain. The 2010 Guangxi wildfire started on March 3, 2010 and lasted 6 days and 22 hours. The fire burnt 93 hectares of forest and brush in the mountain region.

In the article China Pilots Wildfire Detection Sensor Network, we see that China is on the leading edge of fire suppression abilities. We can see that China is advancing its wildfire suppression capabilities which will surely lead to even higher accumulations of fuel, same as in the Western United States and Australia. Eventually as we see over and over again, nature will eventually burn fire ecosystems with fires that will overwhelm the best fire suppression capabilities. All fire suppression does is make for larger more catastrophic fires that are not good for nature or man.

Yet even in this article it is belatedly acknowledged by the Chinese that fire has a role in nature even while the next sentences sound right out of the Smokey the Bear playbook. ☺

"While forests play an important role in global climate and environment, wildfire is common occurrence in China, especially in dry winter, challenges the environmental balance and causes losses in lives and property. According to data release by China's State Forestry Administration, around 2 per cent of the country's forest area, or 28 times the landmass of Hong Kong, is destroyed by wildfire every year." [67]

Currently China relies on manned lookout towers and from forest visitors to detect fire. Satellite imagery systems can detect large fires but not fires in the early stages. The new method is to fill the forests with wireless sensors that can detect changes in humidity and temperature, gas concentrate and the infrared level of the specific area where wildfire breaks out. The sensors notify the control center in 0.5 seconds and the fire scene can be immediately located and a drone sent to investigate.

According to the article Exploring Forest Management, Fire Suppression and Environmental Conservation in China the Chinese are learning the hard way as are there counterparts, that fire suppression leads to more and more catastrophic fire.

They are also learning to appreciate the native people's intimate knowledge of fire being handed down through generations and thousands of years. Fire ecologist Meg Krawchuk traveled to China to see for herself the fire landscapes of China. [68]

"Dr. Shu's data showed that roughly 884,400 hectares of forest burn every year in China. Since 1988, changes in law, law enforcement and fire management have caused the average area burned to decrease by over 90% compared to the 38-year period prior to the Black Dragon Fire of 1987. The Black Dragon Fire burned through roughly 1.2 million hectares in northern Heilongjiang Province, killing 200 people. For comparison, roughly 250,000 hectares burned in the Greater Yellowstone Region in 1988. There is some concern the strong effect of fire suppression is leading to high fuel loading in some areas, as we've seen through the suppression of fire in low- and moderate-severity fire environments in parts of the western United States. Prescribed burning is used in some places as a preventative tool to reduce fire intensity."

"The People's Republic of China contains regions of many terrestrial biomes, and from that diversity one would expect fire would find a niche as an important disturbance agent in some. For example, Heilongjiang Province at the northeastern tip of China largely contains temperate coniferous and boreal/taiga forests. In other parts of the world, fire is a dominant ecological process in these forest types. Though the country is heavily influenced by the East-Asian monsoon, with wet summers and relatively dry winters, lightning during the warm summer season starts many fires in the north during periods of dryness or drought, and there is a growing recognition of fire's importance to these ecosystems."

"In the south, Yunnan Province spans a vast gradient in ecosystems, from the subtropical south to temperate mixed and conifer forests to higher elevation shrub-lands in the Tibetan Plateau in the north. Yunnan is one of the three provinces with the most fire activity in China, but the majority of it is human-caused."

Meg Krawchuk also points out that:

"For at least three centuries, swidden agriculture had been the practice of Yi and Lisu peoples on the hills in the regions around Lijiang. Families would cut and burn patches of forest to clear it for agriculture, then would move to a new area after three to 10 years as soil quality diminished." These old patterns going back hundreds but most like tens of thousands of years have now been stopped with a 1998 law, but questions now remain how adapted to man's activities have these ecosystems become and is stopping this activity wise.

In addition other forms of more extensive agriculture used to grow crops could be more environmentally damaging. "Could it be that some of the biodiversity in these landscapes could depend on fire? Recent controversy over swidden agriculture has suggested that in some situations, the practice can result in a benefit for biodiversity, cultural heritage, soil and water conservation, and carbon sequestration."

Prescribed burning, as Chinese official policy, started in the 1980s after the 1987 Black Dragon fire forced the authorities to rethink their fire suppression position. So in China as well as other areas around the world, some good does come out of these extremely destructive man caused catastrophic fires. Let's hope that there will be more appreciation of the "fire folk-wisdom" of native populations in China that is beginning to develop elsewhere around the world.

Man and Fire in Southeast Asia

In the Cenozoic when India collided with Eurasia, a series of complex chain reactions caused the formation and destruction of sedimentary basins within the domain of the collision belt. This area continues to be very active geologically to this day. Changes in the rate and angle of the convergence between the India and Eurasia plates caused complex kinds of tectonic surface features in Southeast Asia. This created the South China Sea and continuing motions of the Sino-Burma-Thailand, Malay Peninsula, Sumatra and Kalimantan plate blocks produced basins stretching from north Sumatra to central Thailand and on to the Natural area. [69]

All this geologic activity created a diverse ecological landscape in Southeast Asia. The more arid regions are fire ecosystems, but the wet tropical areas are not well adapted to fire. The native peoples have used fire for grazing and agriculture in harmony with nature until recent times. The slash and burn agricultural practices on relatively small areas created a mosaic of plant and animal species diversity.

However this is being replaced by very rapid economic development and ecosystem devastation since the 1960s. Small farmers and native peoples are being thrown off their land by powerful economic interests. International timber corporations go into the tropical forests to log the old growth timber. People then use the logging roads to claim land for agriculture using fire to clean up the logged over landscape. Then other international corporations come in and use catastrophic fire after clear-cutting to clear huge areas for their timber plantations and palm oil plantations. Some displaced people retaliate by burning out planted timber in the timber plantations.

The environmental devastation came to worldwide attention during droughts where tropical rainforests already deluded of old growth timber burned readily. Millions of hectares have burned in Borneo, the Philippines, Sumatra and mainland Southeast Asia. In 1982-83 three million hectares burned on the island of Borneo alone. In 1997 catastrophic blazes set mostly by man produced so much smoke that it created a haze for months that disrupted travel and the air pollution was so severe that it affected 80 million people with a financial loss of over one billion dollars. While wildfire gets the blame, the real culprits are government-linked companies, engaged in clearing the forest to establish palm oil and rubber plantations.

In the book Communities and Forest Management in Southeast Asia by Mark Poffenberger the author makes the point:

> "Fire has always been an important element in managing the natural forest environments of Southeast Asia, especially in the drier ecosystems. Most long-rotation swidden farming systems rely on fire to clear fields, recycle nutrients, and manage pests. Most farmers, with the oversight of community institutions, carefully control such burns. In these contexts, fire has been used for generations as a means of managing the landscape in culturally prescribed ways. Unfortunately, documentation of indigenous systems of fire use and management is limited, vague and judgmental. Some scientists are urging that a greater effort be made to understand the indigenous use of fire and collaboration with local communities."

Mark hits the nail on the head when he points out the real cause for these devastating fires. Rather than blame fire as so many in the region still do, he indicates that he knows better. This is reminiscent of the days of the timber barons in the economically developing United States in the late 1800s and early 1900s. They clear-cut the old growth timber devastating forests and combined with the removal of the native Indians and settlers, caused huge fuel accumulations, that blew up in great conflagrations burning both forests and cities to the ground. In a very similar knee jerk reaction, politicians, economic interests, and even scientists, still blame fire and native peoples for these catastrophic fires and divert public attention away from the real culprits.

Mark is not fooled; he understands who the real culprits are and says:

"It is clear that commercial resource exploitation through logging and mining, as well as estate crop establishment has greatly contributed to the degradation of vast areas, both in terms of their ecological and productive functions. At the same time, over the past century, the role of communities in managing natural forests has been curtailed legally and administratively. A serious political commitment will be required from national leaders and international organizations if the community forest management paradigm to be meaningful empowered through legislative and operational actions throughout the region."

Man and Fire in Europe

The Alps and the Carpathian mountains in Southern Europe were created during the Cenozoic. The climate of the early Cenozoic was much warmer than it is today with a cooling trend, that continues to this day into the current interglacial period. Glaciation began in Europe and North America recently between 3 and 2.5 million years ago and had a huge impact on Europe and American ecosystems and geology. [70]

The great diversity of plant communities today is from the expansion and adaptive radiation of the angiosperms that began in the Late Cretaceous. As the climate cooled during the Cenozoic, plants became more and more specialized with deciduous angiosperms developing in the colder regions and evergreen types evolving in the tropics and subtropics.

The Cenozoic is known to be the age of mammals and led to a mega-fauna of ice age mammals in Europe and elsewhere. Around 8,000 to 10,000 there was a major extinction event that resulted in the disappearance of much of the mega-fauna that has been attributed to climate change and especially the Paleolithic humans with a rapidly improving hunting technology and techniques.

The Neanderthals that occupied Europe for hundreds of thousands of years had learned to use fire as a tool; so they

81

must have added to the frequency of fire in their European ranges. Neanderthals were a lot smarter than folks previously believed and could well have used fire to scare game to stampeded large game into traps. The Article Neanderthals Were Nifty at Controlling Fire points out that Neanderthals were quite skilled in their hunting techniques.

"But the archaeological record shows Neanderthals drove herds of big game animals into dead-end ravines and ambushed them, as evidenced by repeatedly used kill sites -- a sign of long-term planning and coordination among hunters, she said."

As it turns out Neanderthals also had more manufacturing skills than previously thought.

"According to Villa, one of the most spectacular uses of fire by Neanderthals was in the production of a sticky liquid called pitch from the bark of birch trees that was used by Neanderthals to haft, or fit wooden shafts on, stone tools. Since the only way to create pitch from the trees is to burn bark peels in the absence of air, archaeologists surmise Neanderthals dug holes in the ground, inserted birch bark peels, lit them and covered the hole tightly with stones to block incoming air.
"This means Neanderthals were not only able to use naturally occurring adhesive gums as part of their daily lives, they were actually able to manufacture their own," Villa said. "For those who say Neanderthals did not have elevated mental capacities, I think this is good evidence to the contrary." [70]

While Neanderthals were plenty smart, their numbers were not that large, and it seems that it was not until Homo sapiens moved into Europe in large numbers that there was a significant increase in man caused fires. Light intensity fires were first used for hunting-gathering, and later in the Holocene, to remove or open up the European forests into savannas and grasslands for grazing domesticated livestock and agriculture.

Modern human hunter-gatherers began arriving in Europe 45,000 years ago and replaced and or integrated Neanderthals into their genome. The hunter-gatherers experienced varied climatic changes until after the last ice age 11,000 years ago, but in a couple thousand years farmers began to move into Europe coexisting but not initially interbreeding with the hunter-gatherers according to genetic studies.

> "The study identifies the Carpathian Basin as the origin for early Central European farmers. "It seems that farmers of the Linearbandkeramik culture immigrated from what is modern day Hungary around 7,500 years ago into Central Europe, initially without mixing with local hunter gatherers," says Barbara Bramanti, first author of the study. "This is surprising, because there were cultural contacts between the locals and the immigrants, but, it appears, no genetic exchange of women." [71]

Farming that affected Europe was first developed in the high diverse fire environments of the Fertile Crescent also known at the cradle of civilization because farming gave rise to cities and civilization there. The Fertile Crescent is a relatively moist fertile crescent-shaped region in arid and semi-arid Western Asia, in the Nile valley and delta of Northeast Africa. The fire adapted grasses that grew wild were harvested and eaten by hunter-gatherers living in the region allowing for a gradual domestication of these wild grasses over time. Wikipedia states:

> "The Fertile Crescent had many diverse climate, and major climatic changes encouraging the evolution of many "r" type annual plants, which produce more edible seeds than "K" type perennial plants. The region's dramatic variety of elevation gave rise to many species of edible plants for early experiments in cultivation. Most importantly, the Fertile Crescent was home to the eight Neolithic founder

crops important in early agriculture (i.e. wild progenitors to emmer wheat, einkorn, barley, flax, chick pea, pea, lentil, bitter vetch), and four of the five most important species of domesticated animals—cows, goats, sheep, and pigs—and the fifth species, the horse, lived nearby." [72]

The introduction of farming to Europe must have changed hunting-gathering fire lifestyles to fire farming lifestyles and so changed the type of controlled burning these people were doing previous to farming. Burning was surely used to increase the grasslands for livestock, and the introduction of slash and burn agriculture must have had a great effect on forest and savanna ecosystems. The slash and burn agriculture seems to then have been replaced by more permanent agriculture using animals to plow the fields.

Fire regimes in Europe have a wide range from subtropical to arid to boreal. In these fire ecosystems, plants and animals have adapted first to natural fires usually started by lightning and later to more frequent light burning by man. Just like elsewhere in the world the fire ecosystems and all their diversity we have today have been highly impacted by man. If we were to try to revert back to a time before man, that would be catastrophic for many species of plants and animals adapted to more frequent fire. Perhaps it would even be more catastrophic than current fire suppression activity by man.

Fire ecologists classify a large region of the world as Europe and divide it into sub-regions. The Mediterranean region surrounds the Mediterranean Sea, including Northern Africa, and Western, Eastern and Northern Europe. The European fire region extends from Iceland to Morocco including around 50 countries. [73]

Lexi Krock's The World on Fire article does a good of summarizing the role of fire on all continents including Europe.

"The two European fire regions differ in their use of prescribed fire to thin potential wildfire fuels. In the Mediterranean region, particularly Italy, Portugal, Spain,

Greece, and North African and Near Eastern countries almost never use prescribed burning, because it has proved less costly for these nations to fight fires as they arise. On average, some 50,000 fires burn annually in the Mediterranean basin, usually in the spring and summer months. Most of these fires result from human negligence and accidents.

In Western, Eastern, and Northern Europe, by contrast, forest managers increasingly use prescribed fire as a means of both maintaining forest ecosystems for wildlife and preventing the encroachment of bush vegetation. Many of the fires that burn during the summer months in these regions are the result of such prescribed burning.

The largest area of forests in the European region, spanning eleven time zones, is Russia. Approximately 95 percent of Russian forests lie in the boreal zone, as in Canada, and Russia's fire seasons closely resemble Canada's, with most burning occurring in late summer. A majority of Russia's boreal forests are remote and contain large amounts of accumulated fuel matter.

Unmanaged fires burn freely during the fire season. Agricultural burning occurs in pockets of the Russian Federation, and many of these human-set fires quickly become uncontrolled wildfires. In recent years, fire-management teams in Russia have begun to assess ways to limit uncontrolled fires in the boreal forests, for they are a major source of carbon emissions; the carbon stored in these boreal forests accounts for approximately 37 percent of the total global carbon pool."

Man and Fire in the Americas

When South America detached from Antarctica, the climate cooled significantly because the Antarctic Circumpolar Current brought cool deep Antarctic water to the surface. When South America became attached to North America, it continued to cool due to the strengthening of the Humboldt and Gulf Stream currents eventually leading to the glaciations of the Quaternary ice age and to the current interglacial period of the Holocene today. [74]

In the Cenozoic, fire type grasses became widespread opening up many forests into grassland savanna and insects and flowering plants became co-dependent. Grasses played a very important role in this era, shaping the evolution of birds and mammals that fed on grass and the predators that fed on the grazing mammals. Man himself evolved in these fire grasslands and savannas of Africa and then spread around the globe arriving late in North and South America perhaps in several migrations 15,000 to 30,000 years ago from Asia via Alaska and maybe even Europe via Iceland and Greenland.

The Americas stretch almost from pole to pole and with many geological features that created diverse and complex habitats for plants and animals from artic conditions to tropical conditions. In the Cenozoic, the Andes in South America and the Rockies in North America formed as tectonic plates collided.

In the Holocene when modern man spread from Africa into North and South America, he drastically increased the already frequent fire activity caused by lightning. Fortunately during the past century Fire Ecologists have done a lot of research in the Americans on native peoples and fire. We also have the early historical records of white settlers when the first came to the Americas and this really helps the understanding of what the ecosystems looked like prior to European colonization and fire suppression activities.

Today we have this growing awareness among fire ecologists as to the Native American's role in shaping the fire ecosystems of the Americas especially North America. We see this growing awareness indicated in the Wikipedia entry Native American use of fire. When I was a boy Herb Stoddard, Leon Neel, Roy Komarek and my father Ed Komarek Sr. and their associates were well aware of the native peoples fire use before European colonization. [75]

However, today there has been so much more research so that American and Australian Fire Ecologists are leading the way in pointing out the importance of native people's fire practices in these countries and around the world. All I can say is it's about time, being that it was the United States Government that did so much damage to American global ecosystems by promoting extremely misguided and false fire exclusion policies.

This false meme that fire and early indigenous people were bad for the environment is still being supported by bureaucrats, politicians and special interests all around the world closely linked to the fire suppression industry. This false meme is still causing extreme environmental and societal damage in the Western United States and elsewhere, even today, as evidenced by huge catastrophic fires where fire suppression has allowed huge fuel accumulations to build.

In another chapter I will give an overview on the battle for fire's place in global ecosystems that was centered on the evolving Fire Ecologists living in the plantation country around Thomasville Georgia. I grew up in and among these Ecologists, so I had a front row seat to this battle against the US Forest Service and Smokey the Bear. It's a battle that is far from won and is still playing out around the world. It can only be won by widespread public awareness of fire and native man's important role in fashioning global ecosystems.

Wikipedia states:

"Many people believe that North America, before the coming of the Spanish explorers, missionaries, and settlers, was a totally pristine, natural, wilderness world with ancient forests covering the landscapes. This ideal world was populated by millions of Indian people who, somewhat amazingly, "were transparent in the landscape, living as natural elements of the ecosphere. Their world, the New World of Columbus, was a world of barely perceptible human disturbance. This peaceful, mythic, magical ideal — sometimes referred to as tabula rasa — has symbolized the thinking behind much of the modern environmental movement. However, these impressions of a "benign people treading lightly on the land", is wrong in its view of an entirely natural landscape: natives played a large role in maintaining the diversity of their ecosystems.

Fire scientists and ecologists often find old fire scars in trees going back hundreds of years. Geographers studying lake sediments often find evidence of charcoal layers going back thousands of years, attributing the data to prehistoric fires caused by climatic warming and drying conditions. Since the trees and sediments cannot document how the fires started, lightning becomes the easiest "natural" explanation. Early researchers thought that no large burning was carried out by natives, but research during the latter half of the 20th century has shown that many or most of the pre-settlement fires were intentionally caused.

Keeping large areas of forest and mountains free of undergrowth and small trees was just one of many reasons for using fire in ecosystems. Intentional burning has greatly modified landscapes across the continent in many subtle ways that have often been interpreted as natural by the early explorers, trappers, and settlers. Many research scientists who study pre-settlement forest and savanna fire evidence tend to attribute most prehistoric fires as being caused by lightning (natural) rather than by humans. This problem arises because there was no systematic record keeping of these fire events. Thus the interaction of people and ecosystems is down played or ignored, which often leads to the conclusion that people are a problem in "natural" ecosystems rather than the primary force in their development.

Romantic and primitivist writers such as William Henry Hudson, Longfellow, Francis Parkman, and Thoreau were major inventors of the t he Pristine Myth, which became part of American heritage. Influenced by Western prejudice against primitivism and hunter-gatherer societies, many people still believe that Native Americans lived in complete harmony with the environment and neither disturbed nor destroyed but took only what was absolutely needed for survival. One of the powerful technologies which Native Americans had was fire, and they clearly changed the landscape with it. Sometimes to clear the woods, sometimes to create a berry patch, the changes spread across the continents."

I could not have said it better myself. ☺ Fire historian Steve Pyne echoes these sentiments when he wrote:

"The modification of the American continent by fire at the hands of "Indigenous people" was the result of repeated, controlled, surface burns on a cycle of one to three years, broken by occasional holocausts from escape fires and periodic conflagrations during times of drought. Even under ideal circumstances, accidents occurred: signal fires escaped and campfires spread, with the result that valuable range was untimely scorched, buffalo driven away, and villages threatened. Burned corpses on the prairie were far from rare.

So extensive were the cumulative effects of these modifications that it may be said that the general consequence of the Indian occupation of the New World was to replace forested land with grassland or savanna, or, where the forest persisted, to open it up and free it from underbrush. Most of the impenetrable woods encountered by explorers were in bogs or swamps from which fire was excluded; naturally drained landscape was nearly everywhere burned. Conversely, almost wherever the European went, forests followed. The Great American Forest may be more a product of settlement than a victim of it."

It was in the fire grasslands of Mexico that grasses were being harvested for food as elsewhere in the Americas. A very fire adapted grass called Tripsacum that can be found in the Americas was first thought to be one of the ancestors of corn. It has a very tough outer covering, but it can be pounded or popped and eaten like popcorn. The important thing about

Tripsacum is that because of this hard covering it can be stored for long periods without being damaged by insects. It could be an important resource for native peoples when other food resources would be hard to find. It tastes like corn and it was once thought to be an ancestor of corn, but modern genetic testing indicates that corn derived from Teosinte (another fire grass that is found in Mexico).

Corn was developed from Teosinte according to genetic testing from a variety called Balsas Teosinte native to the Balsas River valley in Mexico's southwestern highlands. However, archaeobotanical studies indicate development in the Balsas River valley where stone milling tools with maize residue have been found dating to 8,700 years ago. An early corn was being grown in Southern Mexico, Central America and Northern South America 7,000 years ago.

About 2500 BC, maize cultivation spread over the Americas. It was first cultivated in the United States in 2100 BC. About 1100 AD the size of the cobs and corn expanded greatly from the small primitive varieties being cultivated before. Over a thousand years ago maize finally reached the Southeastern United States after already spreading over much of North America. [76]

Before the White Man, just along the Eastern Seaboard of the United States, there were millions of Indians cultivating Maize and other crops significantly altering the landscape with crops growing in a patchwork of fields among open annually burned forests and savannas. The only areas that did not get annual or frequent burning, was where the Indians kept back fire to grow berry bushes and nut trees. But even these areas had to be rejuvenated over time by fire.

In the Southeastern United States where I live, they would plant a new field either in a small clearing caused by lighting and bugs in the upland Longleaf Pine Savannas they managed with fire, or girdled a few trees using a stone axe. When the soil nutrients ran out, they would move to another clearing or open up one. In order to increase the production on these sites, beans were planted next to the corn to nitrify the soil and squash to suppress the weeds between the plants. When the first Spaniards arrived in North Florida, they found that the road from St. Augustine to Tallahassee was boarded on both sides by Indian farming as far as the eye could see, there was such a high population of Indians in the area.

The Longleaf would then seed into the new field just as it did where there were lightning strikes and so piecemeal regenerate itself. This patchwork of fields also added wildlife cover and successionary stages of undergrowth until the pines grew tall and burnt out this undergrowth leaving wiregrass, legumes and wild orchids for the wildlife dependent on these fire ecosystems. This was so unlike the clear-cutting of the old growth longleaf for timber and cotton plantations that followed in the wake of European colonization and the general extermination of the Native Americans by disease, war and displacement to reservations.

When the Civil War bankrupted the South, the Cotton plantations began to revert back to nature with second growth pines seeding into the old fields. Wealthy people from the North began to buy up this cheap land and create hunting plantations using fire to protect the timber and improve the land for wildlife. Their plantation managers often came from settler stock, who in turn, had learned to manage the land with fire for wildlife from the Indians. The well managed Quail Plantations came to look much like the land before the Europeans arrived and like the way the Indians had managed for thousands of years.

In the central United States, in what came to be called the Great Plains, the Indians managed and hunted with fire to greatly expand these vast tall-grass and short-grass prairies that nature had previously managed with lightning. In the Cretaceous Period, the Great Plains was covered by a shallow inland sea called the Western Interior Seaway. During the late Cretaceous 65-55 million years ago the seaway began to recede

91

leaving behind a flat plain covered with thick marine deposits. [77] Wikipedia states:

> " Paleontological finds in the area have yielded bones of woolly mammoths, saber toothed tigers and other ancient animals,[6] as well as dozens of other mega-fauna (large animals over 100 lb. (45 kg)) – such as giant sloths, horses, mastodons, and American lion – that dominated the area of the ancient Great Plains for millions of years. The vast majority of these animals went extinct in North America around 13,000 years ago during the end of the Pleistocene."

The first Americans arrived on the Great Plains around 10,000 years ago with waves of migration that swept into the North American Continent from Asia across the Bering Straits land bridge. They hunted the plentiful bison as part of the culture of the Plains Indians. The Plains Indians migrated onto the plains from different areas of North America and greatly increased with the arrival of the horse for transportation thanks to the early Spanish explorers. Francisco Vazquez de Coronado, a Spanish conquistador, provided the first recorded account of a meeting with the Plains Indians in Texas, Kansas and Nebraska in 1540-1542. About that same time, Hernando de Soto crossed into what is now Oklahoma and Texas.

It is believed that a combination of frequent fire and buffalo grazing created the mostly treeless Plains in the center of the continent and the forested savannas around the periphery from the relatively recent uplifted Rocky Mountains to the west and the old eroded Appalachian Mountains to east. European trappers moved onto the plains in the next 100 years and by the time of the Louisiana Purchase, one half to two thirds of the Plains Indians had died of European diseases.

During the 1800s, the Great Plains became settled with the Westward Expansion by the Europeans. The remaining Indians were defeated and placed on reservations. Most of the buffalo were exterminated. The overgrazing by cattle farmers, agriculture plowing under of the sod, clear-cutting the forest savannas into the Rockies, had by the 1930s, almost completely wiped out the previous Great Plains ecosystem.

Much of this European caused activity including clear-cutting of the old growth light fire Ponderosa forests, redwoods and sequoias, the overgrazing the grasses on the ground with cattle, and removal of the Indian frequent fire managers, caused huge buildups of fuel. This led to catastrophic fires that by the early 1900s consumed much of the remaining forest and even whole towns both east and west of the Rockies. Government, short sighted public land managers and foresters wrongly blamed fire for the disasters, as they still do in the West today and in developing countries.

These government agencies then instituted misguided fire suppression policies that went into effect all over the United States creating catastrophic fire consequences for generations to come. Even as suppression improved, fuels continued to build through the 1900s and into the 2000s creating a powerful fire suppression industry.

Today this industry has a powerful political lobbying arm that obstructs a move away from fire suppression to frequent fire management. The fledging public fire management teams have little political clout but they are beginning to gain some support from large insurance companies that have incurred large losses from these catastrophic fires.

Ultimately, resistance to these fire exclusion policies began in the 1920s and became centered in the Southeastern United States to which as a boy I had a front row seat in the 1950s and 1960s. I will cover this battle to bring fire back into the Nation's and ultimately the world's forests, grasslands and Savannas, in a later chapter of this book.

In the Western United States, where the prairie leaves off and the land continues on through the Rocky Mountains to the Pacific, we not only find Ponderosa Pine but remnants of ancient frequent fire and catastrophic fire forests, savannas and grasslands. The conifer tree species of the subfamily

Sequoioideae were widespread in the northern hemisphere beginning in the Jurassic. Fossil remains of the genus Sequoia have been found in North America, Greenland, and the Eurasian continent. This indicates that these vast frequent fire adapted forests covered much of northern hemisphere. Only three species have survived the ice ages of the Cenozoic, the Giant Sequoia, Coast Redwood and the Dawn Redwood in Southwest China. [78]

These are huge extremely fire adapted trees that burn out the competition with frequent fire but have been destroyed by misguided fire exclusion policies that allow catastrophic fire type trees to build up in the understory. When these catastrophic fire type trees burn, the fire runs up the trunk into the crown, and the whole tree is consumed in an inferno. Even the redwoods and the sequoia with its up to two feet thick bark cannot withstand these crown fires. They also can't take burning smoldering duff around the roots, where unburned dead organic matter has been allowed to build up over decades.

Conservation measures that took hold in the early part of the 20th century, including the creation of the Serra Club, gained protection from logging in Parks and National Forests and helped save these forests. However fire exclusion policies and not so competent attempts at controlled burning still endanger these forests. It's a daunting task to just learn how to use controlled fire well, let along remove the decade's accumulation of fuel on the forest floor.

In the Southeast, I and others have to burn Longleaf, Loblolly and Slash Pine right after a rain to remove the "duff" one layer at a time for several years until we are free to control burn normally. Inexperienced people either burn when its dry and kill the trees with crown fire, or really hot ground fires, as is still happening today on public lands, or they use a back fire or a night fire to burn through the forest but the "duff" accumulated around the base of the tree smolders like a fuse killing the tree around the base.

In the northern part of North America in Alaska and Canada we run into the catastrophic fire type boreal forests. Some of these trees can take a light burn. In Alaska and Canada the central Black Spruce and willow bog forests and upland Birch and Aspen burn in the heat of summer killing everything to the ground. Over the years through successionary stages, willow sprouts and grasses provide grazing for big game animals like moose and elk and small game like grouse. Eventually the brush, trees and other organic matter builds up and the forest and bogs burn clearing the way for diverse successionary stages to again create habitats for plants and wildlife.

I moved to Alaska to go to College and worked for Alaska Dept. of Fish and Game during the summers and the interior of Alaska would fill up with smoke from the burning fires that would continue into the fall. One of my good friends was a Smoke Jumper who jumped on these fires to put some of them out, but already by this time in the late 1960s and 1970s foresters were beginning to understand that these fires were important for regeneration and in spite of a general fire suppression policy they let fires burn. When the wildfires were no threat to settlements, they were allowed to burn, and besides, they just did not have the resources to put all the fires down anyway. [79]

Recognition of fire's role in fire ecosystems is beginning to advance rapidly in Canada. Many people don't realize that the Great Plains Prairie Ecosystem reaches up into Canada where some of it is being preserved. While the United States has lost most of its prairie to agriculture with some efforts being made to restore it using fire and bison, in Canada they still have some virgin prairie left. Much of it is in Canada's Grasslands National Park (a mixed-grass prairie ecosystem). [80]

In Grasslands National Park, they now use prescribed fire and wildlife grazing to maintain this prairie ecosystem. Because of human development fires can no longer be allowed to run wild so the best that can be done is to simulate wild-land fire with prescribed fire. The objective for using controlled burning in Grasslands is to shift the composition of plants from a non-native to native community, enhance native plant seed production, attract large grazers and reduce the amount of fuel loads in the park. [81]

Much of Canada is forested with boreal forests as is Alaska, but it has much more timber to be managed as a public resource for income and jobs. I remember that when I first traveled through Canada with my family as a boy there was large scale clear-cutting going on devastating the forest ecosystems. Later, when as a young man I traveled through Canada, I saw that there was much more environmental awareness and the timber cutting was being logged in smaller patches, burned and replanted, simulating a natural mosaic of catastrophic fire and regeneration. This is a much more environmental friendly way to derive income from the forests but retain the diversity of plant and animal species.

In the publication, the use of prescribed fire in the management of Canada's forested lands, they state: [82]

"Present uses of prescribed fire in Canada are reviewed. Fire has been a natural component of many forested North American landscapes for millennia, making it an obvious choice as an effective forest management tool. It can be used in harmony with known fire adaptations of ecosystems to be managed. Prescribed fire uses are separated into six categories: (1) hazard reduction which evolved into (2) silviculture (including fire use for site preparation, managing competing vegetation, stand conversion, and stand rehabilitation (3) wildlife habitat enhancement (4) range burning (5) insect and disease control (6) conservation of natural ecosystems."

In Central and South America, fire plays an important role in nature's ecosystems as well providing a livelihood for the people past and present. Because of population pressures and corporate exploitation, what we hear most about is the use of slash and burn agriculture to decimate the tropical ecosystems of Central and South America just as we are seeing elsewhere in developing countries around the world.

It all seems to be fitting into a pattern of resource exploitation beginning in Europe with the removal of the forests for grazing and agriculture, then in the United States and now all around the world as nations move from being developing countries to developed countries. Once the country develops then the people begin to appreciate what has been lost and begin conservation and restoration projects as developed countries.

It is easy for people in the developed countries that are sucking up all these developing countries resources for their higher standard of living to decry the environmental destruction going on in South America and the rest of the developing world. But the protests ring hollow and hypocritical when the consumer of the resources attempts to shift all the blame to the developing world for the environmental destruction.

Yes it is true, the developing countries can learn not only about economic development from us, but perhaps if we are humble enough, they can also learn from our mistakes. There have been huge environmental consequences in Europe and in the United States that are having to be rectified today. If they follow our reckless path of environmental destruction in the promotion of economic development they will also have to deal with the consequences eventually as we are doing.

Fire plays a major role in both South America and Central America. Lexi Krock in in his article the World on Fire states: [83]

South American forests burn each year from both human-caused fires and natural wildfires. Throughout history, humans have practiced intentional burning in South America as a means of land conversion—to prepare land for crops or grazing and to clear large tracts of otherwise impenetrable forests for travel and hunting. Today, it is often difficult to discern which fires in South

97

America are human-caused and which natural. Studies have shown that perhaps 50 to 90 percent of uncontrolled wildfires began as agricultural or land-conversion burns and then grew out of control. In general, as in Africa, the vast majority of forest fires in Brazil are begun intentionally.

Many of the large-scale fires in South America are concentrated in Brazil, Argentina, Bolivia, and Venezuela, where farmers and cattle ranchers undertake prescribed burns. These burns are usually set in and around grassland and savanna environments during the dry season from May to October, which closely corresponds to southern Africa's dry season.

An alarming proportion of South America's burning each year occurs in the Amazon rainforest region, often called the "lungs of the world." Though tropical thunderstorms in the rainforests preclude the ignition of fires by lightning almost 100 percent of the time, farmers converting large areas of Brazilian rainforest try year-round to start burns and keep them going wherever and whenever they can. Unfortunately, cleared rainforest land is rarely sustainable and reverts to non-arable land within a few growing seasons, causing farmers to undertake new burns every couple of years.

In the Chaco Region of Northwestern Argentina are fire grassland and savanna ecosystems. Vegetation in the Chaco Region is composed of a mixture of savannas, thorn shrub-lands and hardwood forests alternating in belts and patches. Rainfall is mostly in the summer months and the winters are dry and chilly with temperatures dropping below freezing. Before man, these grasslands and savannas were maintained by large native grazers and by natural lightning caused fires. When the native peoples arrived, they used the Chaco Grasslands that occupy about a third of Argentina for hunting and foraging and like elsewhere around the globe likely increased the frequency of light fires. Today the fire cycle is estimated to be about 3-5 years. The name Chaco means, "A site for hunting". [84]

Cattle ranches existed since the very beginning of Spanish colonization and displaced or incorporated the native peoples in the area moving them from hunting gathering to agriculture to cattle ranching. Today the area is managed by the activities of ranchers and farmers and fire is used as a tool to push back brush and increase grazing and to keep the land clear for agriculture. The winter of 1993 was a severe fire season as it was very dry and cold. Around 100,000 hectares were burned by wildfire in the Southwestern Chaco Region and in the south 50,000 hectares were burned. This stimulated more interest in Fire Ecology research into the beneficial effects of prescribed fire to better manage these ecosystems.

The article Prescribed Fire Research in the Chaco Region says:

> "Although ranches existed in the region since the very beginning of the Spanish settlement (XVI Century), extensive cow-calf and timber operations began in the mid XIX century largely due to European immigration. The practice of setting fires -- without too much concern about the consequences, as everywhere in the world -- is traditionally used by rangers to promote new growth in early spring, or to 'clean' native ranges. Coupled with overstocking, it led to savanna encroachment by *Acacia, Celtis, Schinus* and other shrubby/spiny species. Severely logged hardwood forest sites have been also invaded by early succession species and kept at this stage by overgrazing."

In Venezuela, it is good to see the appreciation of the indigenous peoples and their fire knowledge. What is also telling is a willingness to cooperate with indigenous people today to develop fire management strategies that are good for both people and the environment. Sletto B. Rodriquez states in the abstract to his paper, Burning, fire prevention and landscape productions among the Pemon, Gran Sabana, Venezuela: toward an intercultural approach to wildland fire management in Neotropical Savannas that: [85]

"Wildland fire management in savanna landscapes increasingly incorporates indigenous knowledge to pursue strategies of controlled, prescriptive burning to control fuel loads. However, such participatory approaches are fraught with challenges because of contrasting views on the role of fire and the practices of prescribed burning between indigenous and state fire managers. Also, indigenous and state systems of knowledge and meanings associated with fire are not monolithic but instead characterized by conflicts and inconsistencies, which require new, communicative strategies in order to develop successful, intercultural approaches to fire management.

This paper is based on long-term research on indigenous Pemon social constructs, rules and regulations regarding fire use, and traditional system of prescribed burning in the Gran Sabana, Venezuela. The authors review factors that act as constraints against successful intercultural fire management in the Gran Sabana, including conflicting perspectives on fire use within state agencies and in indigenous communities, and propose strategies for research and communicative planning to guide future efforts for more participatory and effective fire management."

Man and Fire's Influence on Marine Ecosystems

Best I can tell the ecological dynamics between land and sea is not well understood, let alone how man caused degradation of light fire ecosystems on land might be adversely impacting coastal sea life. One thing is for sure, large amounts of nutrients are flowing into the oceans providing food for sea life. Much of this sea life including, shrimp, crabs, and fish end up on our dinner table as well as on the menu of birds and animals.

Coastal birds and animals then often deposit nutrients back into the uplands and wetlands upstream of creeks and rivers when they roost. For instance on the Gulf Coast, in the Dickerson Bay area, near Panacea about 25 miles south of Tallahassee Florida, cormorants in large numbers feed in the coastal waters and then fly inland to freshwater Otter Lake to roost for the night. One has to wonder how many pounds of nutrients these cormorants alone are depositing into Otter Lake, feeding the lake and its creatures, but also the wetlands and the cypress trees around the lake.

The Ochlocknee River flows down from plantation country in South Georgia and North Florida near where I live and proceeds right on into the Gulf at Ochlocknee Bay, just a few miles from Panacea and Dickerson Bay. At one time, most of the uplands here were a fire climax upland savanna dominated by Longleaf Pine and wiregrass. Due to man's activities, almost all of this has been replaced by other types of pines, hardwoods, farmland and residences.

The wholesale destruction of the once healthy ecosystems in this area has got to be effecting the marine environment on the Gulf of Mexico as well as elsewhere. The kind of vegetation decaying on unburned land must change the kinds of nutrients flowing downstream into the Gulf, as well as farming, logging and even the use of prescribed fire to simulate natural processes in such very fragmented ecosystems.

One wonders if anybody has even done studies on the difference between nutrient decay runoff from different types of vegetation and that of runoff from wildfire and prescribed burns. My friend Jack Rudloe has done a few informal experiments depositing leaves, debris, logs and sticks around what he calls the Living Dock the subject of one of his books.

The idea was to understand if he could simulate natural process of debris floating down rivers into the Gulf to provide a substrate and nutrients for the growth of oysters, a very important commercial fishery in the area. He noticed that most leaves were broken down rather quickly by marine organisms, but not maple leaves for some reason. So we can assume that all types of decay and ash are not equal in nutrients, or how the nutrients are dispersed in the freshwater and marine environment. At this writing, we have been having a lot of rain and the rivers and oceans look like tea full of nutrients for the larval stages of sea life.

I got to thinking about all this and how my father and Jack were at one time starting to focus on this land-marine ecology. It occurs to me that much of the moisture in the form of rainfall we have here in the Southeast, comes off the Gulf and gets dumped into the uplands flushing nutrients down into the Gulf. Just how all this works and the role of fire in all this, should provide fertile research for both marine ecologists and land ecologists working together in the future.

I suspect that one of the reasons this is not well understood is because such collaboration between seemingly divergent perspectives as between Dad and Jack is rare. I think Tall Timbers dropped the ball in this situation with the destruction of Jack's marine specimens held by Tall Timbers post Ed Komarek. I think Tall Timbers and FSU ought to revisit this promising early work and take a leading role in researching healthy upland fire ecosystem effects on marine ecosystems.

Of course there is a lot of work being done studying the interface between uplands and wetlands. While searching the Internet unsuccessfully for references as to fire's effects on marine life, I did run into this study that discusses some interchanges between the New Jersey Pine Barrens maintained by fire being lost to other type ecosystems and effects on marine life. At this point, I expect this is about the best I can do is find indirect references as to fires role in marine ecosystems. [86]

I hope the reader can now better appreciate the importance of fire in nature and also man's past and present role in maintaining these fire environments. Things certainly have come a long way since my father's time.

CHAPTER THREE

THE BASICS OF GOOD FIRE MANAGEMENT PRACTICE

The ability to use fire as tool for management of light fire ecosystems is as much an art as a science and needs plenty of hands on experience. Some of the basics can be taught in the classroom, but much actual field work for students should be supervised by experienced fire management experts. Nothing can really compare to working as an apprentice to an experienced fire manager for years, or even decades. Managing fire is not something that one can BS their way through because much like the stock market or learning to drive, if you don't know what you are doing, you are going to face disaster sooner or later.

There are so many variables to consider when doing prescribed fire, and many of these factors are changing constantly like wind speed and direction, fuel loads and moisture. These factors can cause a light fire to suddenly rage out of control in a few minutes time and turn into a destructive wildfire. There is a step by step process to using fire that starts with a detailed survey of a property, and ends long after the fire has burned through the forest or grassland.

Is the Tract Safe to Burn?

The first thing to do in checking out a potential tract for a controlled burn is to determine if the fire can be easily controlled without too much expense or risk. No point in going further if the fire is going to be difficult to control, unless one has the resources at ones disposal to do it. A good fire manager understands his or her personal limitations and resources as applied to the tract of land to be burned. This rule should apply equally to the backyard burner whose expertise is limited, and whose resources may be nothing more than a rake, a spade and a water hose. This is true for larger projects as well, even for large organizational fire management teams with broad expertise and plenty of heavy equipment.

The safest place to burn is where there are wide natural or manmade unburnable obstacles to fire adjoining the property like plowed farm land, rivers, lakes, highways and property already burned this season. It even helps a lot when one or two sides are protected in this manner because one can adjust how the fire is used to make maximum benefit from such obstacles. For instance, conditions permitting, one can burn away from the weaker fire lines that one has made to contain the fire, and toward the natural obstacles.

Another important thing is to study the flammability of surrounding properties and the existence of buildings on or on other properties. It takes careful care and planning to ensure that no matter what happens, even if the fire gets out of control on the tract, it's not going to burn down somebody's house or barn, on or outside of the property. Fire under the right conditions can throw sparks high into the air and long distances to come down next to a house or barn in dry tinder.

When I was a boy we rarely had to worry about where our smoke went even when it drifted across roads, but now it's a different story. Because of EPA regulations and legal concerns, we often have to put signs up on the road or highway telling people to watch out for smoke. It seems like many city people just don't have enough sense to slow down when the visibility is obscured on the road be it fog, smoke or a mixture of both, so we are required to remind them.

As if there are not enough problems with controlled burning, now we have to try to burn away from the highway, and folks burning large tracts have to be careful not to have the smoke drift over cities else they will be fined by the EPA. Also because of local regulations involved and fire permits, safer late afternoon burning and night burning are becoming ever more difficult, and this earlier in the day burning also can cause heat damage where really light fire is needed.

In the Southeastern United States and in some other regions as well, fires go out late at night because of the dew. In burning in the late afternoon in the Southeastern United States we know that the fire will be slowing down all evening. If the fire should get away or get too intense being started in the morning, it can do a lot more damage before it is contained by the evening dew.

Does the Tract Need Prescribed Fire?

Once we see that the tract of land can be safely burned then we can walk or drive over the property to see if it should or needs to be burned. The first thing that a good fire manager notices is fuel accumulation on the forest floor and up into the brush beneath the trees. The more fuel accumulation the more important it is to burn the property. In most cases if we don't burn under controlled conditions, nature will sooner or later do it under uncontrolled conditions and with disastrous consequences! It's not a case, if the property will burn, but who burns the property and when. ☺

The next thing one studies is the type of vegetation on the property to see if the ecosystem needs regular periodic fire to survive and flourish. In most parts of the world, except for some areas in cold climates and in some tropical landscapes, where light fire ecosystems are predominate. However, there are always fire tender ecosystems surrounded by larger light fire ecosystems. For instance in some areas, holly, beach and magnolias, in the Southeast because of fire exclusion, have moved out of wet almost unburnable areas onto higher ground in the domain of fire species like pines and fire resistant hardwood and softwoods.

Decisions have to be made ahead of lighting the fire as to protection of these out of place sites, unless the overall goal is to restore the property as a fragmented ecosystem to within natural parameters. For instance, we might want to restore the uplands to fire resistant species like pines and grasses and leave the lowlands to fire tender species that naturally should be occurring there with no or very light fire. This can be a very tough call for any fire manager or land manager.

Such special treatment also involves extra resources needed to work against nature, rather than with nature, if we are to protect fire tender species on high ground. However, if one has a beautiful stand of large beautiful magnolias in an upland fire environment, it's going to be tough not to protect them from fire, not only from the controlled fire but also from wildfire. It gets even tougher if one has a beautiful stand of large pines and magnolias because if you don't burn, the pines will eventually die out failing to reproduce.

If you do burn, you kill the magnolias unless you use very light fire just to get the fuel loads under control, maybe even having to rake around individual magnolia and beach trees. Fire management and land management work is going to be difficult for those who don't like making decisions in favor of one ecosystem and against another. These types of people often manage by default because in not making a decision they favor not only catastrophic fire, but do great harm to natural light fire ecosystems.

It's important to remember that a land manager is an artist painting on a living canvas that is changing all the time. The land manager makes decisions based on what he or she wants from the land or what the owner wants. The owner might want to create many diverse ecosystems on the property simulating nature, which might include growing timber for profit, opening up grassland, developing marsh ecosystems on the edges of wetlands for wildlife etc.

If the objective is to develop diversity, I always study the wetlands carefully to see if I can slam hot head fires (fire being driven by the wind) into the edges of wetlands to push back trees and brush to favor marsh, reeds, ferns and other wetland species that flourish in wetland savannas. Wetland savannas also provide habitat of many wildlife species that do not do well anyplace else. Problem is that in some areas like in the Southeastern United States wetland savannas have disappeared and turned into hardwood forest right up to and into the water because of infrequent fire.

I also take a good look at the wetlands themselves, streams, lakes and rivers on the property. Much of our plant and animal species in wetlands need fluctuating water levels and places like swamps need to be burned out in times of drought in order to remove the accumulating peat, else the swamp turns into lowland.

Of course this should only be done when the land around the swamp or lake edge has been burned in wetter times. If you don't, peat fires can burn for months and creep under or through a fire line and set wildfires elsewhere. This is a serious problem in some places in North Florida USA even when burning the woods in dryer times. In places like this, one has to know even the characteristics of the soil upon which one is burning or plowing fire lanes. Dead material in the soil can burn under or through a plowed fire line like a fuse for weeks and end up on the other side of the fire line to start up a wildfire. Peat fires are also really difficult to extinguish, you may think they are put out but several days later they can start right back up again.

Preparing Property for Controlled Burning

Once an evaluation of the property is made, it's time to prepare the land for burning, by plowing or even mowing the fire lines. Usually fire lines are plowed to plow under burnable grass, leaves and needles in an area wide enough to contain the fire. Sometimes folks don't want to plow up the ground so they mow a grassy road really close and then wet the road down using a sprayer just before burning.

If the area is a backyard no more than an acre or two, a fire line can be raked with a hand leaf rake or mowed close with a lawn mower. A water hose with extensions to reach around the whole backyard property can really get an inexperienced homeowner out of a lot of trouble. A water hose can really put out a lot of fire and wet down a fire line. It's also good to cool down brush and leaf piles that start burning too hot, throwing sparks and burning leaves off of the property. Care must be used to locate the hose where it does not catch on fire else it becomes useless, especially when needed the most. ☺

Large landowners, if they don't have their own equipment, can contact the county forest service and for a reasonable fee they will plow the fire lines with a small dozer pulling a plow. If the fire line has never been plowed, this is a really good deal because the dozer has a blade in front that pushes small trees and brush out of the way making way for the line. Usually this first plow is with a sod buster blade that makes a deep trench that can easily erode from a lot of rain, so it's best to have them go back over the line with a harrow to create a flat plowed surface that does not erode easily, especially on downhill slopes.

The first order of business using the plow is to plow a good wide outside fire line all around the property to be burned as a last defense against fire getting out on the neighbor's property. Then one goes about cutting the parcel up in smaller blocks taking advantage of natural or manmade firebreaks already on the property. So if we should have fire get out in one block we can still can hurry back to the main line set a backfire and stop the fire before the head fire jumps the mainline fire break and becomes a wildfire. If one has burned the property a lot and has a lot of experience, one may just use the main firebreak, but it's risky and it's always good to have multiple lines of defense if one or two lines are breached.

Another reason for burning in blocks is that if the fire goes out at night it can still pick back up from a burning log or stump burning like a fuse, then race with the wind even days later, jump the fire line and turn into a wildfire. So it's best to completely burn a block before going on to the next one. One also has to pay special attention to any dead trees that can catch fire, or even dead material in live trees that can blow sparks across the fire line starting a wildfire across the line. A dead tree when upright can burn for days and then fall across the line to burn to the other side like a fuse. The land manager may no longer be checking in the days after a fire. So we want to cut down dead trees before burning, or rake around them so they don't catch fire.

Another thing to consider is that fire can burn over water in sedges and grasses driven by the wind to burn across a marsh or wetland to start a wildfire on the other side. It just takes years to understand all these complex variables and manage them properly when one uses prescribed fire. For native peoples it was no big deal, one just raked around the village and fences and then set fire to the woods and fields. This practice of open range burning even carried over into the 20th century in many places. Now the only place this is possible is on islands surrounded by water.

Within a block one may even plow additional fire lanes. For instance, if we want to slam a head fire into the edge of a wetland, we might set a fire line back from the wetland 50 yards to light a head fire off of the line, so it has time to build up a wave of fire. Or we may simply note that when we burn we will light a head fire into the wetland and then let it back into other areas that we do not want to burn so hot.

This backing a fire into the wind is called a back fire and is usually low and quiet, slowly backing into the wind, unable to build up a wave as with a head fire. Of course the wind can shift and a back fire can turn into a raging head fire, so one always has to consider this possibility and have several fail safes set up just in case this happens. A good fire manager is always aware and on edge anticipating wind shifts, to stay ahead of the game, and not have to play catch-up and even have a fire get away.

So a good fire manager looks over the land very carefully and decides ahead of the time where he or she is going to light fire throughout the block. This once the fire is secure along the inside of the fire lines. One may want to burn through this little briar thicket with a head fire, but back the fire right next to it through some young tender pines etc. In this manner, a great amount of micro-diversity can be created even on a small piece of land, if it's not just being burned on a government industrial scale for fire control with little or no attention to ecological niches.

Lighting the Fire and Keeping it Under Control

In order to pick the right time to burn we keep a close eye on the weather as that makes a big difference as to how well and how hot the fire will burn. If we have a thirty year rough in Longleaf Pine, we may even burn several hours after a rain to take just one layer of debris off the top of the thirty year debris accumulation every year for several years. We also do this so that the dead bark falling around the base of the tree does not smolder and burn like a fuse girdling the base of the tree.

Normally in the pine lands of the Southeastern United States that are burned annually, we like to burn three or four days after a rain in winter after the needles and grasses have dried out sufficiently from the sun and from frosts. In Longleaf Pine wiregrass upland savannas, we may even burn the day after the rain because the dead debris is so flammable and we don't want to unnecessarily scar the base of the trees or singe the needles up in the canopy. Longleaf Pine has evolved very flammable needles and thick insulating bark to burn out the competition.

Usually the second day after a front comes through with a rain there is a steady north wind, so we don't have to worry so much about wind shifts that begin to happen on the third day after a front. Unless we really need to burn brush or oak leaves, we avoid burning when it gets really dry because with shifting winds and even high winds, the burning gets hazardous and very tricky even for a very experienced burner.

On pine land we also want to burn in winter when it's cool and before the new needle buds sprout new needles. In the South this is after the middle of December and before the first of April. Because the Quail Plantations in the area where I live rely on timber for some income and Bob White Quail production, the plantations burn after the quail season ends in late February and before the new buds on the pines sprout after April first.

March is also a good month to get down on brush that tends to choke up the drains making it difficult to hunt quail. Government burning on an industrial scale often ignores damage to trees and micro-habitats in an effort to just protect the forest from wildfire, as their fire managers are underfunded, the timing of burning is limited and also because most of their monies don't come from timber or quality of habitat but from the taxpayer.

Often those that burn a plantation or a farm have lived on the land for most of their life even generations with intricate land management experience being handed down from parents to their children. This is very much the kind of intimate detailed contact and bonding to the land that we have seen from native peoples before colonization. A good plantation manager pays attention to single trees. He may even pick up a handful of moist dirt to throw on a burning fire scar so the scar

111

does not get worse and multiple fires burn down the tree eventually.

Usually the final decision when to burn in made in the morning of the day to be burned when all conditions for an appropriate controlled burn are in place. On a large tract the equipment is cranked up and positioned, and on a small tract the landowner makes sure the water hose is connected and the rake and shovel out of the shed.

Particular attention is paid to wind direction and velocity. We usually go to the downwind fire line to light a spot fire on the inside of the fire line to begin. We light the backfire along the downwind fire line so that as it grows it makes a wider and wider line to stop the head fire later. If this is not done and a fire is lit off the upwind side of the line, the fire will travel as a wave of fire and burn through the tract and then jump over the downwind fire line.

So the upwind fire or head fire is lit only after all the downwind line has burned back far enough to be secure. When the spot fire is lit, we watch the smoke as it rises to give us indication of wind direction at higher levels and we also see how well the debris is burning. If the wind is not satisfactory or the debris too dry, we might even cancel burning that day. If everything looks okay, we light a small section of the backfire line and then watch again to see how the fire is behaving and so on.

Right now is when a fire manager is the most nervous trying to figure out how the fire is going to behave because no fire is the same as any other because of so many variables. As the fire manager moves along the line, he or she gains in confidence that the fire is behaving properly and with the downwind side of the tract protected by a wide backfire burned area, the manager can breathe easier. Its sort like driving a car or truck you have never driven before, but as one gains experience the car or truck sort of becomes easier to drive and one feels more comfortable.

The other thing is that even a backfire can throw a spark across the plowed fire line, so the manager is going back and forth through the smoke along the back line making sure no spark is thrown across to start a head fire on the other side. If a spark does get thrown to start a spot fire across the line, the manager knows that if not dealt with immediately with water or a rake, it will soon blaze out of control because now the fire can travel with the wind.

When the back line is secure, then the manager proceeds to light the fire using a rake or fire torch along more of the side lines toward the upwind side of the property. He or she proceeds depending on how hot the fire should be. The manager may proceed up one side into the wind with a diagonal fire that moves between cool and hot and only light the head fire on the upwind side as the day progresses. We know that when the debris begins to get damp the head fire will not get too hot as it races through the property with the wind.

It takes quite a bit of experience to work up the line into the wind keeping the fire at just the right intensity to do the job. The manager also knows that once he goes completely around the tract the fire will create its own wind and pull toward the center increasing the intensity of the fire. In pineland, I usually get to this point late in the evening after the sun has set and the fire will not burn good, unless the tract does get encircled. In addition, it's much safer to burn now with all the fire pulling away from the fire lines and moving to the center of the parcel.

When I burned Birdsong, now Birdsong Nature Center, [87] as a young man I developed lots of experience with that piece of property. I could burn the whole outside line in one day in order to make the property secure. The next day I would travel about inside the property using different intensity of fires to develop unique micro habitats all across the property without having to worry about the fire getting away from me.

Detailed Ecosystem Burning

I guess a good example as any on how to burn a tract to develop diversity would be Birdsong Plantation owned by our family until it was donated and became a part of what is now Birdsong Nature Center. It was first opened up in the 1930s with much of the property cleared and put into improved cow pasture by my father with a little help from my mother. He did the burning with me as an apprentice, then turned the burning over to me alone for a few years in the 1970s and early 1980s, and then later to my mother.

By the 1960s Dad got out of the cow business and slowly let the fields revert back into simulated natural ecosystems, while continuing to experiment with fire. He used the place in support of his research on fire and if it worked on Birdsong, he would scale up the idea to be used on Greenwood Plantation that he managed at the time. Greenwood was about 10,000 acres composed of three main parcels to the west of and south of Thomasville, and it also included a smaller piece with a large lake and some very big bream that were delicious.

Like many plantation or farm boys, I grew up developing an intimate knowledge of the parcel of land I was raised and in the case of Birdsong it was 565 acres. By the time I could walk, I was riding on my father's lap on the tractor as he managed and burned Birdsong religiously every single year.

It is important to understand a little of the history of a parcel of land before and during burning. For instance, from pieces of pottery, arrowheads and flint chips, one can see there were three Indian habitation sites on Birdsong with both near seepage springs. Two springs are next to what is now the Big Bay Swamp and another next to the Big Pond. During slavery days a dam was built using a mule pan to hold water for a rice field below that was made very flat. Later my father used equipment to further raise the dam so the swamp grew in size to 65 acres. So we see historically that the contours of the land have been changed by man as well as a large increase in the amount of wetlands on our property.

What my father started doing while I was growing up was to slam hot fires into the marsh around the whole of the Big Bay Swamp in December before the rains came. This would burn out peat accumulations and keep the Button Bush under control. It allowed for open water and marsh and a diversity of marsh and marine life. The shallow water would soon be teeming with small fishes and invertebrates providing food for larger fishes and wildlife such as turtles, alligators, herons and ducks.

Below the Bay Dam and all around the rice field and back up to the Big Pond, my father showed me how to slam hot walls of fire into the edges of what was now a Black Gum shallow flat swamp. This fringe area between the fields and the old shallow rice field built up a wonderful ecosystem of reeds, sedges and trees in what we would call a wetland savanna. I even planted some pitcher plants in this area and they are still there today. The fires would really snap, crackle and pop, when they reached this high fire type area. Over time, these hot fires cut into and pushed back the bay trees creating more savanna.

We would do the same thing around the other ponds that Dad had constructed around Birdsong. Some ponds were wet weather ponds and others held water year around. When we burned through the Bay Field Pond, a wet weather pond, this created an excellent habitat for frogs because fishes could not compete and live there. In dry years the pond would dry up. When I camped next to this pond in spring, the sound of the frogs was so loud one could hardly get to sleep. Usually there would be a small alligator in the pond feeding on the frogs and small turtles. When there was water in the pond in wet years, plenty of ducks and herons could be found feeding in this shallow pond. The endangered Wood Storks would come in when the pond got low before drying up. Large numbers of

Wood Storks arrive when a pond is almost dry because their food is concentrated in a small area and easy to catch.

Another pond we called the Frog Pond would hold water year around, so I put catfish in it and when they overpopulated, I put bream and bass in the pond. Wood Ducks really liked this pond as well, and the Wood Storks, when the water got low in dry years. Around this pond we burned the productive marsh to keep back the trees in this niche. In later years, a leak developed in the dam and in dry years this pond now dries up in dry years.

A wetland is really ideal when it has a fluctuating water level and is burned. When the water is low, grasses and sedges grow on the dry bottom, and when they are flooded they provide food for wildlife. A pond with a stagnant bottom does not have near the diversity in the water and around it that a pond where the water level changes during the year. So fire along with manipulation of water levels really increases the diversity of plant and animal species simulating natural processes in fragmented ecosystems and environmental niches on a parcel of land.

Now, as to the field and savannas on Birdsong, my father and I had our work cut out for us after he took the cattle off the land. We found that without a lot of grazers and just a few browsers like our Whitetail Deer it was difficult to keep our open land and we had to increase the intensity of the fire to compensate. Even this was not enough and over many years we slowly had to face the fact unless we were going to use equipment, as we did in some fields, that the open land was going to turn into upland savanna.

Again this is where knowledge of history really helps because at one time not that long ago there were Eastern Buffalo and wolves, and ten thousand years ago mammoth and mastodon. I assume like the elephants in Africa, mammoth and mastodon were good at pushing down trees, clearing brush, and along with the buffalo allowed for very lush highly combustible fire type grasses to dominate like Tripsacum.

116

Dad had introduced Bahia grass in the fields for the cattle, but when the cattle were off the fields the land began to revert to native grasses and sedges including Broom Sedge. I figure that over hundreds of years, if left alone, we could end up with the uplands returning to a Longleaf-Wiregrass Fire Ecosystem. It would be a typical fire savanna with high fire grasses like Tripsacum in the drains along with non-fire climax forest of hardwoods, beach and magnolia, along the creeks here in the Southeastern United States.

So without these other species of grazers and predators that the white man and the Indians before had wiped out over thousands of years, the overall ecosystem must have been quite different than now, even in areas where we have been simulating natural processes. We just have to do the best we can, but it would be interesting to bring in buffalo into this area and even elephants to see just how much effect these beasts have on fire ecosystems.

A little farther north there are experiments in the Eastern United States involving the reintroduction of elk and some interest in how elk along with fire could reshape man devastated ecosystems. As we have seen with wolves in Yellowstone even the reintroduction of just one species, wolves, has even led to geological changes because of the effect on elk populations and I assume buffalo populations as well.

We have a long way to go to understand the loss of many different species of animals and plants on forests, savannas, wetlands and grasslands. Tripsacum was almost wiped out in the Southeast by cattle as it can't take overgrazing and it's not hard to visualize a mammoth or mastodon along with Saber Toothed Tigers wading through savannas of Tripsacum.

I was responsible for reseeding Tripsacum on Birdsong after my father got to experimenting with it on Tall Timbers Research Station. As a young man not even yet out of high school, I used to scuba dive working for Tall Timbers and FSU in local rivers in Florida to find the remains of these beasts no longer in our ecosystems most likely because of the hand of man.

So you see creating diverse ecosystems is an art as well as a science. We try to experiment to improve diversity of plant and animal species on land, but as can be seen it's not easy, it takes a lot of knowledge and folk wisdom and one really needs this strong bond to the land that only really comes from growing up and living on the land one's whole life.

After the Burn

For the uninitiated stepping out onto the burn just after the fire, the landscape looks like a vision of hell, black and deluded with dead logs, stumps and rat nests, still smoking. The tree trunks have been blackened and the land covered with a coating of ash. The land manager keeps checking the fire lines to make sure no hidden brands have ignited across the fire line and that no burning dead trees have fallen, or are going to fall across the fire lines.

If there is a burning dead tree, we have to take it down with a chain saw, or try to put it out if it's not burning too badly with a water sprayer. This is tricky because even when you think you have the smoldering tree out, it can start burning again hours later from just a tiny ember hidden someplace in the tree. This is why we keep checking the line until it gets damp and then we can go home to rest up because the next morning we need to check the lines again about ten in the morning. This is when the dew begins to evaporate and fire can travel again if a tree or spark has taken the fire across the fire line.

Usually by this time any dead tree that we thought we had out but did not, would be smoking again. In areas where the fire line has cut through dead roots or peat in dry sandy soils one might have to come back to check from time to time for weeks. There are places like this in the Flatwoods south of Tallahassee Florida.

118

On well managed lands like Quail Plantations, the land managers never leave the land and are out most every day doing other work than fire. They study the land that they have burned, watch it green up, and by summer the uninitiated would not even know the area had been burned unless they noticed a fire blackened pine trunk of fire scar. In the subtropics vegetation grows very rapidly due to the heat and rain, and so it is imperative that we burn every year to keep all this vegetation under control, except for areas where we need to allow some plant species to grow two or three years between burning.

Most of the industrial burning by government agencies is three years or more and this may be okay to protect forests from wildfire, but for most species of wildlife it is not good. Not much wild life or vegetation on the ground can survive being tightly overgrown by trees and most wildlife species live off the ecosystems on the forest floor and if shaded out there is not much but leaves and needles. This is another reason why selective cutting (mimicking natural processes of thinning) is not just good for income, but for the wildlife as well, as it opens up the pine canopy so more sunlight gets to the ground.

Over the year the land manager observes his or her handy work and may make changes to the burning based on what is observed and from experiments on the land the next winter. Summer burning in some instances where there is good pine overstory may create additional diversity on a small scale, but by and large the best policy is to winter burn. One most definitely should not summer burn fields without an overstory of pines as my father and I found out fast enough on some plots we developed on Birdsong.

One of the tenants of good fire management is to experiment on a very small scale, then scale up if the experiment is successful and to repair the damage if not successful. It is very time consuming and expensive to bring in heavy equipment to repair damage, or even worse to just ignore the damage that was done that has resulted in a loss of habitat and diversity.

In the case with summer burning on Birdsong's fields, we found this fire was slowed down by green growth and would even creep around the brush allowing the brush to get high enough to stand up to fire and become trees. We did discover how easy it was to herd insects and I even experimented with eating those roasted by the fire, or caught moving ahead of the light slow fires. I found I liked them both raw and cooked.

Now might be a good time to bring in the views of both Herb Stoddard and Leon Neel, both who have integrated and practiced good fire management techniques for both forest and wildlife management on a micro-management scale.

The Stoddard-Neel Approach to Timber Management in the Southeast

Most people are aware of the devastating effects of large scale clear cutting and regeneration on public and private lands that devastates ecosystems much the same way as catastrophic fire. But few are aware that there is a better way to manage and regenerate forest timber and ecosystems on both public and private land. What the Stoddard-Neel Approach does is provide for a continuous regeneration and improvement of a forest, while bringing in much needed income for property owners.

Just as with prescribed fire, we mimic natural process leaving a light footprint on the integrity of the forest and its ecosystems. However it requires an exhaustive understanding of the natural ecosystems being managed over a long period of time and for generations. The forest is kept healthy and intact as the forest is progressively thinned as it grows and is regenerated over long rotational cycles in small parcels.

Of course environmental activists are aware that logging corporations often try to use simulated thinning as a pretext for excessive logging of public lands. We have some problem with this even in private forests when the owners or land managers are not familiar with good timber management practices. We don't want the logger in control over the management of timber because they will often over thin because it's in their best short term interest. We also don't want careless loggers scaring up the bases of pine trees as they pull cut logs out of the forest with their skidders.

So the way it works in plantation country is that we have independent foresters like Herb and Leon who worked directly for the landowner and manager and not for the logging or land clearing companies. The independent foresters charge a fee to oversee timber management which involves marking trees and then following the logging to make sure that only marked trees are cut. They also make sure that trees are not scarred by falling cull trees, or by careless skidder operators pulling logs out of the forest in a hurry. Each individual tree to be culled is marked waist high is also marked with paint on the base, so the forester waking through the timber stand can make sure that only the allotted trees are cut and no more.

Often the independent forester even personally handles the controlled burning or prescribed fire for his or her clients. Over many years working for a client, the forester develops an intimate relationship with the land being managed and works closely with the land manager who also has developed an intimate relationship with the specific parcel of land over decades for both timber production and wildlife production. I think large bureaucratic centrally controlled land management agencies can learn a lot from how decentralized private plantation lands are managed incorporating native "folk wisdom". In The Art of Managing Longleaf it is stated:

> "While Stoddard-Neel is not a formulaic approach to forest management, it does rest on several fundamental commitments. One is to single tree selection. Single tree selection was not a new technique in the forestry profession when Herb Stoddard first started practicing it; its merits had been debated for some time. But in Stoddard and Neel's

iteration, the central proposition was a conservative marking strategy and the long-term survival of the forest, rather than the most efficient production of timber resources.

Each tree taken, in effect, mimicked natural, small-scale disturbance events such as lightning strikes, blow-downs, insect damage, and other natural mortality. Such events open gaps in the forest canopy, allowing more light to hit the forest floor and making room for longleaf regeneration to take place. To open these gaps, Stoddard and Neel first selected the weakest trees for removal, leaving the stronger trees to build up the timber volume so the fewer, but more valuable, trees could be removed the next time around."

As the reader can see, there is a sustainable and economic way to manage ecosystems both with fire and with good management practices that simulates natural wild-lands. But it takes a lot of understanding of these ecosystems to do what is best for nature and for man. There is a middle ground that is good both for man and nature and maintains important diversity of ecosystems by simulating natural processes. The simulating of natural processes has been refined to a high art on Southeast hunting plantations and can be adapted to ecosystems elsewhere. Ponderosa Pine forests for example, in the Western United States are similar in many ways to pine ecosystems in the Southeast. Herb Stoddard had this to say about burning Longleaf Pine in plantation country.

"Within our region there is considerable diversification of flora and topography, and this has a bearing on just how control burning may best be handled. When I say best, that's my own viewpoint. Other people may have better ways of doing it. First: Whenever possible, and in almost all conditions in which we use fire as a tool in land management, the soil should be damp at the time of burning. The duff, and remains of limbs, logs, and other punky material and debris from past selective lumbering operations, should be wet.

If dry, it may smolder for days or weeks, often constituting a hazard for nearby pine areas through "pickup" fires.

Second: Burn in pine lands from high combustibility to low. For example, the combustibility of Longleaf Pine debris under given conditions of wind and moisture is much higher than that for Slash, Loblolly, Shortleaf, and Pond Pine. Set fires first under Longleaf stands, or where this pine predominates on the hilltops, when conditions of dampness and air movement are right for light burning. If wiregrass (Aristia) constitutes most of the ground cover, rather than broomsedge (Andropogon), the fires may burn "just right" soon after a "front" has gone through with an inch or more of rain, followed by sunshine and wind.

A mixture of wiregrass and Longleaf straw traps much air, which accounts, in part, for its extreme inflammability, so it should be burned when so damp that "swingeing" fires will only burn downwind. Some of you may question the word "swingeing," but we've used it for a long time and think it has a real meaning. Such areas may best be burned only a few hours following the rain. Next day the fires may be re-set and burn farther down the slope under the less inflammable Loblolly and Shortleaf Pines; it may take several days of sunshine and re-setting of fire, before we can complete burning through the bottoms, and under scattered stands of pine.

Careless users of fire not infrequently reverse this procedure, starting to burn a unit only when the lower slopes can be made to burn through. The result is likely to be a severe "blow up," when the flames reach the top of the hill covered with highly inflammable Longleaf and wiregrass. That I think should be the "a" in the alphabet of controlled burning. Always burn from spots of high combustibility when conditions first becomes right to spots of lower combustibility when conditions me be right only after several days. Make a mistake and the woods are very unsightly and damage may be done to the Longleaf."

In summary, even though I have used the subtropical Southeastern United States to illustrate how we control burn, these basics of burning can pretty much be used from arid to arboreal landscapes.

CHAPTER FOUR

FIRE WARS – THE BATTLE FOR FIRE IN NATURE

Prelude to Battle

The battle against 20th century total fire suppression in nature's ecosystems, and for the use of controlled fire in nature, was the result of the convergence of two different perspectives on how nature should be managed. The convergence was between a few European educated United States government foresters with little ecological understanding, especially of fire's role in nature, and a few old museum collectors who were evolving into ecologists to become the founders of the new emerging field of ecology. The age of the collectors and taxonomists was ending and these old collectors and a rare academic or two found themselves founding the emerging field of ecology.

Science was evolving rapidly in the 1800s and into the 20th century. This was especially true for those attempting to use the scientific method and principles to understand nature in what was called at the time the Natural Sciences. Collecting and classifying had begun in the Natural Sciences in earnest centuries before, even before the time of Darwin. The 1700s, 1800s and early 1900s were the age of the collectors and taxonomists, because in order to study something like nature, you need to figure out what's there to begin with. This requires collectors going out into the field to collect and preserve specimens and gather information on those plants and animals. It also requires that there be taxonomists in the museums who are trained to name and categorizing specimens in some kind of logical scientific manner.

In a similar manner, foresters were evolving from the 5th century monks who established a plantation of Stone Pine for use as a source of fuel and food in Byzantine Romagna on the Adriatic coast. Forestry practices were developed by the Visigoths in the 7th century when facing a shortage of wood. They passed laws concerned with the preservation of oak and pine forests. Systematic management of forests for sustainable timber began in the 14th century in Nuremberg and in 16th century Japan. Before all this, the Chinese had a long history of forest management. According to Wikipedia:

"Schools of forestry were established beginning in the late 18th century in Hesse, Russia, Austria-Hungary, Sweden, France and elsewhere in Europe. In the late 19th and 20th century forest preservation programs were established in British India, the United States, and Europe. Many foresters were either from continental Europe (like Sir Dietrich Brandis), or educated there (like Gifford Pinchot)."

This is a very important piece of information because it had a very big impact on the creation of United States government forest fire suppression policies in the early part of the 20th century. [88]

The reason that this is so important to understand is that for hundreds of years European forests were being removed to make way for farming and grazing as Europe's population skyrocketed due to advancements in farming and later the industrial revolution. Much of this destruction started as hunter and gathers were being replaced by farmers for thousands of years and it was sped up by advancements in technology and new crops that could support more and more people. It started with rotational slash and burn agriculture that became permanent when the soil could be tilled easily by animals pulling plows.

We can be sure that the emerging modern foresters were horrified in the 1800s at the near total destruction of much of Europe's forests as Europe industrialized, and it was natural to blame the local inhabitants and fire for this destruction. Today this same process is happening in the Amazon, Africa, and Southeast Asia. We can see this same kind of knee jerk reaction here as elsewhere by called "educated people" without ecological training.

When these "educated" foresters without ecological training rose up in government ranks in the early part of the 20th century, they were horrified as they experienced huge catastrophic fires that were devastating the forests in the United States along with great loss of life and property. Like generations of foresters before them, they saw forests going up in smoke right before their eyes repeating in the United States what had already happened before in Europe. But because they did not have ecological training or a proper understanding of fire in nature, they saw fire as the enemy, not the widespread logging and loss of the Native American light fire managers.

Well, they were about to get that ecological training because in the Southeastern United States their new fire exclusion policies ran up against some newly emerging collectors turned ecologists. These very stubborn scientists and land managers associated with the game plantations around Thomasville Georgia knew that this total fire suppression was wrong and very damaging to natural ecosystems and they were not about to roll over even when threatened with jail time by "educated" foresters for their burning the woods.

The so called "educated forestry scientists" were soon to get an education by some real scientists, who over decades proved through scientific studies that the idea that fire was bad for nature was an ecological destructive myth. As it turned out, the government foresters had built their anti-fire propaganda on little or bad science, but it took decades to begin to undo the damage of United States Forest Service propaganda. The job is far from over even today as forests weighed down by huge fuel loads accumulated by the suppression of light cleansing fires blow up into unstoppable catastrophic fires, not only in the Western United States, but in the rest of the world as well.

There was yet another factor that the fire exclusion foresters failed to understand and that was the value of a close relationship with nature as shown by native peoples all over the world. These early "educated" foresters had a real contempt for native peoples and their understanding of fire, and a way to high opinion of themselves, as was commonly held until the middle of the 20th century.

Contrary on what you might view on Wikipedia, the transition and battle for prescribed fire was anything but reasoned and dispassionate. The anti-fire opposition was so full of professional and academic hubris that they were right, and with no scientific studies to back them up, resorted to demonizing and ridiculing anybody who stood in their way. One needs no better example of the mood and tactics of the anti-fire forces even as late as 1940 in regards to scientists and local people alike. Psychologist, United States Forest Service John Shea, writes in 1940 in an article Our Pappies Burnt the Woods: [89]

> "Outsiders visiting or motoring in the South during burning seasons, however, are shocked and appalled by the miles of fire running free in the woodlands and the palls of smoke that dull the sun and often make motoring hazardous. "Why," they demand, "cannot these fires be stopped or controlled?""
>
> "WOODS burnin' 's right. We allus done it. Our pappies burned th' woods and' their pappies afore' em. It war right fer them an' it's right fer us." So spoke a lean resident of the piney woods—one of hundreds I interviewed in the course of a six months' study last year during which as a psychologist I was supposed to find the "inner-most" reason why inhabitants of the forest lands of

the South cling persistently to the custom of burning the woods. "Fires do a heap of good," continued my "patient." "Kill th' boll weevil, snakes, ticks an' bean beetles. Greens up the grass.""

"The extent of the annual burnings, the harm they do and the barrier they raise to successful forest culture throughout the South are well known to federal, state and private forest agencies"

"Seeking a new educational approach, the federal Forest Service last summer decided to delve deeper into the human or social roots of the woods-burning problem. It was hoped that here might be found a point of vaccination that with an improved educational serum would reach the germs of the woods-burning desires."

There was a least one forester that did not cave to anti-fire hysteria in the Forest Service. James Barnett comments: [90]

"Fire protection became a moral crusade and early Forest Service researchers were generally proponents for complete control of fire. However, based on his research with the Urania Lumber Company, H.H. Chapman became a proponent of controlled use of fire as a means of controlling wildfires and, more importantly, stimulating forest regeneration of southern pines. In a 1912 article, Chapman argued that to keep fire entirely out of southern pine lands might result in complete destruction of the forests. Later, in 1926, he issued his famous Yale University School of Forestry Bulletin 16 which caused controversy among southern foresters because he called for the use of fire in longleaf pine regeneration."

"Chapman was a charismatic, but forceful character. He published more than 20 papers between 1909 and the early 1940s dealing with southern pines and their relationship to fire. His work showed that most winter fires do not kill all longleaf pine seedlings; but they helped establish pine stands, suppress pine and other hardwood competitors, and reduce hazardous fuel accumulations." Chapman recommended use of fire in longleaf pine every

128

three years. For his pioneering work, he has been termed the "father of controlled burning for silvicultural purposes".

In a sense, the new ecologists grew up among native peoples as they went about their collecting duties for museums like my Dad and Uncle Roy, or had been brought up in their early years on the land and close to nature as had Herb Stoddard and Leon Neel. Of course don't get me wrong, there were some government foresters who caught on pretty quick, as decades of fire suppression led to huge wildfires in Florida in 1943. By the time I was born in 1948 I fell right into the thick of things. ☺

The Field of Battle

The reader might wonder just why did the battle for fire begin and catch hold in the Southeastern United States, and in particular, the area in Southwest Georgia around Thomasville, when fire exclusion was being implemented across the United States in the 1920s. What was so unique about Southwest Georgia? In order to answer that question I need to provide some history of the area.

The Red Hills of South Georgia and North Florida formed geologically when the sea level was higher and a great river flowed down into the shallow waters of the ocean. This river carried clay erosion down from the Appalachian Mountains, when they were much younger and rugged than today, and deposited it into an alluvial plain under water. When the sea level fell, this clay was eroded into what are called the Red Hills of today. The unique thing about the Red Hills is that this red clay based soil holds water and nutrients unlike the surrounding area of the sandy coastal plain. Trees and vegetation grow much better here than in the local sandy soils where the water and nutrients soon drain away.

This is important because as the cotton farmers before the Civil War used up the land farther north, they migrated south looking for fertile soils in which to build new cotton plantations and they found just what they wanted in the Red Hills. Of course, the Indians that they displaced were using these same fertile soils for at least a thousand years to build up large farming populations of people.

This of course was before they died of disease, when the first Europeans arrived, or were finally displaced by the cotton farmers in the region in the 1820s and 1830s. Previous to this, Jackson had invaded the Southeast with his Cherokee allies pressuring a melding of what was left of several different tribes called the Seminoles out of the area. Jackson's men brought back word about the good soils available to the cotton farmers back north.

When the South lost the Civil War, most of these cotton farmers were bankrupted in the depression that followed the Civil War. The railroad traveling down the eastern seaboard had been extended and ended in Thomasville, Georgia. Wealthy northern businessmen and their families had begun working their way south in the winters looking for recreation and hunting in the South to escape the northern winters. They found that hunting quail and other game in the Southeast was as sporting as could be found in Europe.

First these wealthy northerners stayed in Thomasville hotels and just rented land to hunt from the local farmers whose lands were falling out of cultivation in the depression. These lands were naturally seeding into second growth pine-grassland savannas. The farmers that maintained these old fields with fire on an annual burning cycle had created just the habitat for the best Bobwhite Quail hunting.

By the late 1800s the wealthy elite business families from the north were buying up the land around Thomasville building homes or remodeling the old plantation houses that survived Sherman's march through the South. They put the old farms together into plantation hunting lands managed for both wildlife and timber. They discovered that a well-managed pine forest, selectively cut over a hundred year rotation simulating natural thinning, could provide income. Additionally the simulated thinning served to continuously improve the stands of timber, by culling out the weak damaged trees, leaving only the tallest straightest trees.

Often the businessmen hired as the farm managers the people who before them had owned and managed the land. In such a manner they had people with lifelong intimate knowledge and experience, living on and managing the land with fire. When the local fire manager died or became disabled a son might even take over having been trained in management as an apprentice.

In January of 1894, Henry L. Beadel came to Tallahassee with his father and brother. In his article in the First Annual Tall Timbers Fire Proceedings, Henry recounts his first experiences in the area. His father's intent was to shoot quail and Henry's purpose was to shoot quail and also get some schooling. At that time Tallahassee's population was only 2500-3000 people. Henry and his father and brother hunted in an area about 10 miles north of town with the help of a two horse power hunting wagon and a black driver named Charley. His first experience with fire was in late February when it seems the whole of the land was on fire and he and his brother became quite upset until Charley explained what was going on.

> "Nothing in its aspect suggested to us that the land had ever been burned off. So our ingrained northern notions about fire suffered a shock when one day toward the end of the quail season, in late February, we saw the whole country on fire, which within a few minutes left the ground black and bare except for scattered clumps of bushes. The country looked to us irretrievably ruined, and the quail doomed, until our trusty Charley informed us that this burning took place regularly every spring as far back as his great-grandpapa could remember. Our alarm thus somewhat abated when a few calmer squints through the smoke showed all the trees still standing, and we even

found that we could walk behind the flames without scorching our boots.

In 1895, Henry's uncle Edward Beadel bought Tall Timbers Plantation on Lake Iamonia about 15 miles north of Tallahassee. Henry says that during his uncle's ownership controlled burning continued. From the period between 1897 – 1912 Henry was not able to come hunting but in 1913 onward he came and hunted with his uncle every winter until he took over the plantation in 1919 and made it his home and carried out the burning for another few years.

In his Tall Timbers Conference paper, Henry gives a rare account of the contribution to burning done by black workers on his and other plantations in the area. All too often, it has been the whites that got credit for burning the woods and fields of the old south, but much of the time the burning was left to the black slaves during slavery times and as sharecroppers, after they were freed.

It should be kept in mind that at this time there were many freed slaves still alive. It also should be noticed that like in Indian times there were no firebreaks and everybody just got out and burned their area unconcerned about the fire burning adjoining properties.

"On the last day of the quail season our head Negro "made a narration" to the tenants giving them permission to "put out the fire," which, the buildings and fences having already been raked around, they promptly did, "putting the fire" to everything that would burn. That night, on every hand, lines of flames crept or raced across fields, flickered through pine woods, here and there flaring high over the heaver clumps of weeds, accompanied by cracklings of brush, bangs like pistol shots, and clouds of eye and nose-stinging smoke."

Those were the first night fires I had seen close up or so to speak from the middle, and they were vastly more impressive than day fires. Fascinating spectacles they were. Yet they conveyed the sense of menace which any wildfire, though known to be merely skimming the soil harmlessly, is apt to inspire, particularly in one like me, who has camped and hunted in northern woods where some soils unlike ours are vulnerable to fire, and where any fire not under absolute control can cause widespread destruction of both soil and timber. Anyway the sight of those night fires increased my dislike for burning more than was strictly necessary and led to speculation as to whether we still needed to burn as often (that was yearly) as the early settlers found to be best; or whether, under present conditions, we could get along with less frequent burning without seriously affecting the ecology."

So in order to settle the question Henry set us some experiments plowing fire lines around his property and into sections. Henry says by this time 1926, Fire Control, was primarily for preventing fire from invading adjoining property and had become fairly standard practice. The frequency-of-burning experiments were just to keep fire out of plots from one to three years. He says three years of exclusion was enough to convince him that the old time settlers knew very well, in fact better than people like him do now, what they were about when the burned yearly.

The Battle Begins

So we can see things were proceeding along just fine until the early 1920s when the Forest Service anti-fire propaganda machine rolled into town and convinced some of the plantation owners to make their managers stop using fire. Henry Beadel was smart because he innately knew one of the major precepts that applies to fire and other experimentation as well, start small, and if the experiment works scale up.

In Henry's case he saved himself a lot of trouble and equipment expense by doing his fire experiments on a small scale first. Other less savvy plantation owners stopped their burning altogether with serious consequences for years to come to get their properties "back-in-shape". It only took two or three years for the land to start growing up in brush under the pines where it was increasingly difficult to drive their hunting wagons across the land, let alone see the dogs. Even worse, the quail disappeared almost overnight being adapted to annual fires.

Most of these wealthy owners were not stupid, and quickly realized their error and allowed burning again, but they realized that if there was no scientific work backing up the need for fire, the government would soon force them to quit burning and they would lose the hunting value of their plantation. This is a clear case of accountability in the organizational structure of the plantations. However, government bureaucracies with little oversight can get away with putting off the consequences of fire suppression. They can increase fire suppression for decades with little accountability from the public.

Powerful and wealthy men such John Hay Whitney were not about to roll over and lose their winter recreation easily. They came together and they soon devised a plan to finance a scientific study of the Bobwhite Quail that would scientifically prove that frequent fire was beneficial to the Bobwhite Quail and other game. This study would serve as a foundation helping to protect their investment in their forests and game from wildfire devastation. It was a good thing they did, because the government propaganda operations were moving into the area big time with the Dixie Crusaders.

In Col. Thompson's long house hunting lodge, plantation owners and their girlfriends could escape their wives, under the manly hunting pretense. In front of the large blazing fireplace here on Meridian Road, on cold winter evenings, the Southwest Georgia and North Florida plantation owners would conspire to bring Herb Stoddard, a taxidermist and field collector, to scientifically prove why their hunting population of Bobwhite Quail had collapsed. Scientific evidence they knew would undermine the powerful Smokey the Bear propaganda operations of the Forest Service soon to give rise to the 1928 invasion of the Dixie Crusaders into the Deep South. Rooney writes: [91]

> "A fleet of special trucks-equipped with generators, (many of the hamlets visited lacked electricity) and motion-picture projectors, and manned by articulate young southern foresters-headed for the woods in September 1928. Between then and June 1931, the Dixie Crusaders, as they came to be known, preached the gospel of fire prevention to three million people in Florida, Georgia, Mississippi, and South Carolina."

After these informal meetings between plantation owners at Sherwood Plantation, a formal meeting was held at the Links Club in New York City. Following this meeting negotiations were begun between the Committee of the Cooperative Quail Investigation and the chief of the Biological Survey, and an agreement signed on February 5, 1924. Field headquarters were to be Col. L.S. Thompson's hunting plantation on Meridian Road where Herb was invited to stay and run the investigation.

Herb Stoddard's scientific, rigorous and expansive book The Bobwhite Quail was the final report of the Committee that began work on March 17, 1924 and ended June 30, 1929 and was published in 1932. In this exhaustive book is a watered down chapter on fire. The book was only published after Herb threatened to publish the book himself, if he was forced to dilute the fire chapter any further by the Bureau of Biological Survey, United States Department of Agriculture.

This quail book firmly established Herb Stoddard as a founder in the newly emerging field of ecology and fire ecology. Herb was a good friend of another more well know father of ecology, Aldo Leopold, the author of The Sand County Almanac. A third founder of the ecological field was Dr. Alee who mentored my Dad while he studied at the

University of Chicago before Dad went to work for Herb in the early 1930s.

Aldo Leopold was aware of the importance of fire where he lived and wrote his classic book the Sand County Almanac in the prairie country of the north central United States. He also recognized as did Herb the importance of native peoples fire knowledge. In a paper in 1924 he stated:

> "Previous to the settlement of the country, fires started by lightning and Indians kept the brush thin, kept the juniper and other woodland species decimated, and gave the grass the upper hand with respect to the possession of the soil. In spite of periodic fires, this grass prevented erosion. Then came the settlers with their great herds of stock. These ranges had never been grazed and the grazed them to death, thus removing the grass and automatically checking the possibilities of widespread fires. The removal of the grass relieved the brush species of root competition and of fire damage and thereby caused them to spread and "take the country."

At the end of the paper that Herb Stoddard gave at the First Annual Tall Timbers Fire Ecology Conference called Some Techniques of Controlled Burning in the Deep Southeast, he said:

> "I know a paper like this may be just the "a" of the alphabet to many of those in attendance, but I thought a little reiteration doesn't hurt when you're talking about fire; we don't want to take any chances with a tool as dangerous as fire. Almost all tools that are effective in land handling can do great damage if wrongly used. An inexperienced operator with a bulldozer, in a young stand of pine, can do just about as much damage as a wildfire. I'm inclined to agree with some others here that we can learn a great deal from primitive man if we are humble enough."

In 1943 came the revolution in U.S. Forest Service thinking in Florida after decades of fire suppression, even when some foresters and landowners were becoming very alarmed by the buildup of fuel in their forests making them increasingly prone to wildfire. Even while Smokey the Bear anti-fire proponents were still firmly in control of government fire policy, there was rebellion in the ranks. R. A. Bonninghausen, Chief of Forest Management of the Florida Forest Service, writes in the 1962 First Annual Fire Conference Proceedings:

> "In 1943 "came the revolution. Prolonged drought throughout the south coastal plain area brought about the most disastrous fires that any of us had experienced in the history of protection in the South. I think the biggest fire was on the Osceola National Forest. As I recall, it was 70,000 to 80,000 acres. It was a tremendous thing. I was working for a paper company near Gainesville as resident manager and a pall of smoke hung over that area for days. We sat sleepless, worrying about our own piece of timber because things were so dry and the accumulation of rough was so great in many parts of the forest. In our Gulf County unit, which is a Florida Forest Service protective unit, I believe 20,000 acres burned over in one fire.
>
> In that same year there was a conference called at Lake City by the Forest Farmer's Association and it was a conference on fire. To the best of my knowledge that was the first one that was called in this area concerned principally with a discussion by foresters as well as people who were intimately acquainted with the use of fire in the woods. At that conference, in looking over an old program which I dug out of the files, I found these names: H.H. Chapman, H.L. Stoddard, Frank Heyward, John Currey, Claude Bickford, Hux Coulter and many others that we recognize today as having had a great deal of influence on these changes."

Herb Stoddard, one of the major players in the pro-fire wars family, moved to Florida in 1893 and lived there until 1900 where he developed an early understanding of the importance of fire from the local settlers. He then settled in Chicago where he worked as a taxidermist for the Chicago Field Museum. As it turns out another major player in the pro-fire wars, my father Ed Komarek Sr., came from Chicago where his father had emigrated at the turn of the century and started a vegetable business and later got into construction.

With the great depression hit, Dad's father was wiped out and he had to quit college where he was being mentored by Dr. Alee. Dad grew up in a suburb of Chicago that still had some wildland and as a boy he had a strong interest in nature especially birds. He had soon ingratiated himself to the taxonomists at the Chicago Field Museum that worked down in the basement.

One of these sniffed this white power and Dad learned later that this friend was the doctor for Al Capone. It must have been here that he was taught to skin and stuff small birds and mammals with cotton to be put into museum trays for study. I suspect he was also doing some collecting for the taxonomists in the local area because when he had to quit college he landed a job with the Field Museum for 50 dollars a month and the use of a vehicle to go collecting mammals in what was to become the Great Smokey Mountain National Park. Later, as things got even worse in Chicago where Dad saw bread lines, his mother asked that he take his younger brother under his wing and get him out of town. They were to work together as a team for the rest of their lives.

At the time Dad had no contact with Herb. Herb showed me as a boy some of his pictures when he was involved in cleaning out and mounting rhinos and elephants by digging into the insides of these huge animals and carving out the meat and bones to leave the hide for mounting. It was only later that Dad heard about Herb and his work where he lived on Meridian Road in Southwest Georgia. Dad first met Herb on a small mammal collecting trip south to Florida and he and Herb became good friends. The brothers went on down into Florida where Roy stayed for a while living and working closely with the Seminole Indians in what was then a fire savanna grassland occupied and burned by cattlemen.

Dad then went to work for Herb as his assistant in the Cooperative Quail Association study in the 1930s. After being mentored by Herb on fire he was soon converted and enthusiastically embraced controlled burning or prescribed fire as it is now called. Ed begins his Fire Ecology paper of 1962 with this line:

"In presenting this paper on Fire Ecology I find myself in somewhat of a quandary for I have spent about half of my life influenced, taught and educated against fire in nature and then I have spent the other half of it using fire and trying to understand it."

By the 1940s, Dad quit his low paying work as Herb's assistant, did some consulting, and then went to work for the multi-millionaire John H. Whitney to manage his 10,000 acre quail plantation near Thomasville, Georgia. Leon Neel took over from Ed, and Herb and Leon went into the timber consulting business together after the Quail Association ended. Dad told me that he got the job at Greenwood in the following manner. (Leon was to become another of the founders of Tall Timbers Research Inc. and a player in the pro-fire wars up until the time he left the board a few years ago to write his book on the Art of Prescribed Fire. He is the only one still alive today of the original founders of Tall Timbers Research.)

Major Beard, who was managing Greenwood at the time, decided to look Dad up and traveled to Birdsong Plantation. It happened that Major Beard was struggling to modernize and make the transition from mule power to modern equipment power on Greenwood. When Major Beard finally found Dad, he was a sight to behold as he emerging in spectacular fashion out of a thicket of brush 15 feet high. Ed was driving his iron spiked wheeled tractor pulling a heavy drum brush cutter that was dwarfed by the brush he was chopping down. This was how Ed, along with fire and some bulldozer work cleared our Birdsong Plantation's fields to raise cattle and timber.

If I recall correctly, Dad was offered the Greenwood job on the spot by the very impressed Major Beard who was having brush problems on Greenwood. Dad finally began to make enough money to build up the family's Birdsong Plantation next to Sherwood Plantation to its final size of 565 acres. Even more important he had an agreement with Mr. Whitney that his scientific work would be considered part of his regular work at Greenwood as long as things were operating smoothly.

Soon Dad brought in his brother Roy to work at Greenwood under this same agreement and both after a few years had things running so well that they could devote most of their time to fire research. Dad with the backing of Roy, had the ability, the financial resources and time to organize the first fire conferences filled with speakers from all over the world as well as pull together the founders to organize Tall Timbers itself in the late 1950's.

I was born in 1948 and by the age of five I was already being mentored by my father and Herb in the use of fire. I learned the basics very early and at this age was sitting on my father's lap learning to drive the tractors and the family's 1947 Willy's jeep, one of the first civilian jeeps in the area. I was brought up around the top ecological scientists of the day and learned by example. I was laying fire with a rake at least as far back as I can remember helping my father burn the woods.

One of the ways I had a front row seat in the fire wars debates was at the weekly meetings on Sunday that started around ten in the morning in our home's living room before and after mother had the big plate glass bird window put in. Herb was the first to show up for morning coffee and popcorn served by my mother and sometimes by my sister and me.

I was usually watching out and greeted him as he was like a grandfather to me teaching me how to skin and mount birds and he even gave me his climbing spurs as he aged. He always walked with a limp from a tree climbing accident where he fell out of a tree most likely using those same climbing spurs that severely damaged his knee. It almost happened to me a couple of times too. ☺

Next Roy, my father, Sonny Stoddard Herb's son, Leon Neel and anybody else that happened to be around showed up. The land management discussions would get very heated, so much so that my mother would walk out of the house upset. Then soon afterward they would be best of friends again and could not understand why my mother was getting upset as they were just blowing off steam as they put it. ☺

I understand that the same heated discussions carried over to the Tall Timbers Board meetings. In a sense these informal meetings every Sunday morning at Birdsong, became de facto board meetings before and after Tall Timbers was organized. I never sat in on Tall Timbers board meetings as a kid but I always looked forward to Sunday and to participate in the lively Sunday morning discussions. Everybody else felt the same way who attended.

Roy lived first in Thomasville and later moved onto Birdsong and from the very beginning was like a second father. My sister and I always ran out and greeted him hugging his legs just as we did with our Dad and Mother. In fact Roy threw out his back for several weeks playing horse with me and my sister on the living room floor.

Developments in ecology and other topics that were going on that week were usually on the agenda of these friendly gatherings. Ed would often come up with an idea that Roy being the conservative of the pair would try to shoot down, only to have Ed go along with the idea anyhow and Roy would fall in behind to make sure the idea was implemented properly.

There was another critical player in the early days of the pro-fire debates and that was "Cowboy Stevens" who lived on his plantation near Tifton Georgia about two hour's drive away. Our family would go to Tifton once a year to be with the Steven's family and stay at their cottage on the plantation.

Mr. Stevens, best I can recall, spent a lot of time in the 1920s collecting fire grasses for the Tifton Experiment Station to improve cattle grazing in the South. In fact, my father planted the cow fields on Birdsong with Argentina Bahia grass that really improved cattle production in the South. Importantly, Mr. Stevens spend a lot of time in Africa and told Dad a lot about fire in Africa and had pro-fire contacts that Dad later brought to speak at the Fire Conferences.

By the 1940s and 1950s, Dad found his true calling in the struggle to stop Smokey the Bear propaganda and misguided government fire suppression policies. In the late 1950s, he organized and was one of the founders of Tall Timbers Research Inc. His intent was to undermine U.S. Forest Service propaganda with facts and to use the organization to distribute that information not just in the United States but around the world.

By the late 1950's Herb was getting into his later years and had a timber consulting business to run, but was very involved with the bird kills on the WCTV television tower and the fire plots on Tall Timbers. Herb agreed to be president of the new organization to bring in funds because he was so well respected by plantation owners for his lifetime of work in the area.

Leon Neel was also a founder of Tall Timbers and continued to serve on the board after Ed and Roy were gone. As of this writing, Leon is 90 years old by still going strong and I have had him critique this book for suggestions and to correct any errors of fact I may have made.

However, the real power and most of the work in the early years at Tall Timbers was done for free by Ed and Roy with Dad firmly in control as the institution's executive secretary. He pretty much used the institution and Roy to turn his ideas into reality and to fight Smokey the Bear tooth and nail, not just locally but all over the world. ☺

I guess I should point out that Ed and Sonny Stoddard had a kind of sibling rivalry, because in the early years Ed and Herb had become so close. They could not be in the same room with each other without acting like a couple of roosters joking around trying to get the better of each other. Later in life Sonny criticized Ed for his salesmanship, but Ed's charisma and salesmanship allowed him to promote fire and Tall Timbers in a beneficial way. He used his TV program Rural Report to promote seed corn along with his ecological initiatives locally.

Ed also used his salesmanship and his passionate pro-fire activism to extend Tall Timber's reach globally, making Tall Timbers under his leadership, a global organization. Ed also developed creative ways of salesmanship as is often the case when one does not have enough funds to promote ones work and efforts.

Mother claimed she came up with the idea to give away small bags of the new Dixie 18 seed corn to the agriculture 4H groups, but Dad made good use of this idea. When the sons and daughters of the farmers planted this free corn their parents saw the advantages of Dixie 18 right away, and began ordering large amounts of seed corn from Greenwood.

Dixie 18 was very successful in putting Greenwood Seed corn on the map in the south and was very profitable to begin with. The seed is still around as I have heard from a local farmer friend who planted some recently. The Company eventually went out of business when the national seed companies began to develop specific regional strains.

This kind of creative salesmanship and promotion also spilled over to the fight against Smokey the Bear propaganda operations. Ed made the first Tall Timbers Fire Conference Proceedings free worldwide and this immediately put Tall Timbers on the map as being the leader in fire ecology research, lasting till the end of Dad's tenure at Tall Timbers.

As pointed out by Jack Rudloe, Ed like Herb before him mentored many people. In fact, I have taken a chapter out of Ed's playbook by making my activist books free on the Internet to reach the largest possible audience. While Dad used snail mail and world travel before the Internet to organize and promote, I do the same with the Internet and the social networks today. This has been very necessary for my work just as making the first Proceedings free were for Ed and Tall Timbers. When one has little money, one is forced to become very efficient and to think outside of the box.

I used to think as I was growing up that Dad and Roy were obsessed with fire, only to learn later in life that it takes this incredible focus of attention, passion and personal resources to change human consciousness in an extensive way. This small group of men and women independently, financially backed by savvy plantation owners, found themselves on a mission to restore fire to its rightful place in the global environment. The battle to put light fire back into light fire ecosystems is far from over and I hope this book will make a small contribution to that battle by following up on my father and his associates' work.

Dad and Herb like many naturalists and ecologists were quite spiritual in their own way. To Dad the tall Longleaf pines at Greenwood Plantation where he worked were the pillars of nature's cathedral. Because of this spiritual connection to the natural world, he really felt very passionately on a deep visceral and personal level, the incredible damage being visited on fire environments all over the world, by government misguided fire exclusion policies. Passion, rather than paycheck, is a very important aspect of all kinds of activism. Passion drives a person to overcome obstacles that paychecks cannot do.

Ecology and Fire Ecology evolved out of the natural sciences where many of the field collectors like Ed, his brother Roy and Herb noticed that creatures were in dynamic interaction with each other. The transition was not easy for many, especially the museum taxonomists who sorted and catalogued museum specimens. Dad used to argue with these men in the basement of the Chicago Field Museum. They would say there is no basis for ecology because there is no such thing as a relationship that cannot be touched, smelled, tasted or seen.

In fact the early ecologists had a pretty hard go of it. Most people thought it down right strange and quirky to not just watch birds and mammals in the early part of the 20th century, but to skin and stuff them with cotton to be sent to museums! Many of my friends think I am quirky to have an interest in the paranormal and wonder how I became that way. They have no further to look than my scientific background. ☺ Those who follow the path of truth worry little what others think of them.

Think what it must have been like to the average person who might have met Ed and Roy on muddy Meridian Road as they traveled to meet Herb in the late 20s or early 1930s. The Model T or Model A Ford collecting vehicle would be stuffed with the tools of the collecting trade. In the late twenties and early thirties, roads were mostly unpaved, muddy, slippery and rutted so the vehicle most likely was covered with mud as it slips and slides down muddy Meridian Road after a heavy afternoon shower.

A collector had to often sleep outdoors and live off the land. Inside the vehicle might be found some usual items such as canvas tent, sleeping bags, tarps, food supplies and cooking utensils, warm clothes, matches, flashlights and the ever useful shovel and axe. Scattered amongst the usual items were the more unusual tools of the collecting trade. There would be rat traps, mouse traps, mole traps, pocket gopher traps, steel traps, collecting guns, pistols, snakes in cloth bags, butterfly nets and poison jars to kill the butterflies without damaging their wings. There would be jars of formaldehyde and alcohol full of the insides of critters that contained a variety of unusual parasites, fetuses and worms.

One can well imagine the variety of sights and smells that such strange cargo held. Perhaps the most interesting and intriguing would be a bucketful of maggot looking creatures that crawled and squirmed around in the sawdust in the bottom of the bucket. This seething mass of creatures could be seen busily eating away the flesh on two or three mammal skulls that had been dropped into the bucket several days before.

When the dermastids were through with their grizzly task, the skulls would be clean and ready to ship north to the Chicago Museum of Natural History. Sent along with the skins to the museum would be the skulls of the hapless creatures to which the skins had once belonged.

Other collecting tools would be the Eastman Kodak Auto Graflex box camera enclosed in its alligator hide case. When in use, the top opened up into a pyramid where one peered down into a mirror and through the lens while preparing to take pictures. The film came in plates that would be peeled then pushed down into the back of the camera. Cameras were important to collectors because they could be used to record the habitat of the creatures being collected. My father also carried an over-under Gamegitter collecting pistol manufactured by Marble Safety Axle Company. This versatile long barreled pistol had a 22 rifle barrel on top and a 44 shotgun barrel below. There was also a 410 shotgun pistol that was also easy to carry in a backpack. The Gamegitter was really good because depending on the critter and the situation the damage to the skin could be limited by which barrel he shot the animal with.

The skins from a large animal would usually be sent back to the Chicago Museum of Natural History as a flat skin unless it was for a taxidermy display. The skin would be slit open on the belly from head to tail and front leg to front leg and hind leg to hind leg. It was then peeled off the body. The skin was scraped and dusted for preservation, then tacked out on a wall or a board to dry. This was also the manner in which fur trappers preserved and shipped their skins for hundreds of years. The preservation of the smaller mammal skins was different.

Think what it must have been like for a farmer or hunter happening upon Ed and Roy encamped by a stream with both working on small portable tables to skin and stuff specimens. ☺ A small tray was the platform used for this type of work and it was set on the lap, or on some small table while working.

In the tray were an assortment of surgeon's tools such as scalpels, tweezers, clippers, scissors, along with cotton and wire.

Borax was liberally scattered all over the bottom of the tray to soak up blood to protect the fur or feathers of the skin. A small slit would be made with a surgeon's scalpel on the belly of the animal's body and the creature would be skinned through this small opening in the belly. The skin was then peeled back and rolled inside out in borax or arsenic for preservation, then turned right side in.

Finally, the critter's skin with the fur or feathers side out would be stuffed with cotton and wire, then stitched up to resemble the live animal or bird with maybe a stick sticking out of the specimen's behind. With birds this was particularly true as the specimen could be picked up and rotated without touching or messing up the feathers. The specimens would be stored away to dry in trays in a special box carried in the back of the collector's car until ready for shipment. One could well have observed close by, the carcass of the animal or bird if it was eatable, roasting over the campfire for supper. ☺

In this strange and curious way, the field of ecology slowly evolved out of the natural sciences, mentored by passionate, quirky, individualistic people desiring to understand nature and natural processes. These were people who were close to nature, as were farmers, hunters and fishermen, but who had at least some scientific training either formally or were self-educated.

Only later would academia make claim to the field as its own, and flesh out in greater detail the ecological concepts of fire research and management, but often without the passion that drives change. The problem with academia is that it is an indoctrination process whose structure evolved out of the Church, but with today's scientific dogma. It is no accident that academia still has much of the structure, indoctrination and ritual of the Roman Catholic Church.

Academia still uses Latin names for plants and animals and is constructed autocratically around courses, diplomas, degrees

147

and graduation all part of priest class society going even further back in time to Egypt and Sumer. It is also true, in spite of what is insinuated, that academia no more owns science, than the Church owned religion in centuries past. Much of cutting edge science that is of a generalist nature was developed by non-academics, and only later fleshed out by academic specialists.

In fact, academia can be quite toxic to the generalist because of the specialized nature on the indoctrination process. Fortunately, these free thinkers called by the Latin name heretics who rejected Church dogma, are no longer burned at the stake, but are still degraded, marginalized and snubbed by ivory academics of today's scientific establishment. No wonder exceptional free thinkers like Jack Rudloe had such violent early confrontations with Academia

Dad had to be tough and thick skinned because the opposition from government agencies was really great. I remember once Dad even threatened to contact his Senators and Congressmen when the United States Forest Service would not even allow one of their members to come speak at a fire conference with the way being paid for by Tall Timbers.

It's obvious that by the 1930s and 1940s the battle for American and global ecosystems was fast heating up with men like Herb Stoddard using scientific research and truth against government power and propaganda operations. By the time I had grown up in the 1970s, the truth was out in the public domain, and the importance of fire in nature scientifically known all over the world.

Still, it must be understood that to get to even where we are today in fire management, it took not only scientific research on fire, but strong fire activism to combat Smokey the Bear. The battle still wages in the Western United States and Australia showing that even huge amounts of science and truth are not enough to overcome the inertial of decades of fire suppression propaganda and special interest money in public land bureaucracies and political offices.

How and Why Did Tall Timbers Drop the Ball?

We can see from the increasing catastrophic fires around the globe today that fire research is not enough and fire activism is required. Tall Timbers was organized by my Father not just to do research, but to involve itself in fire activism. When he ran the organization, he and the other founders used Tall Timbers as a tool for fire activism as well as research. He made sure that the organization was funded primarily with private funds so that public institutions could not attack its funding when they came under fire for poor fire management.

The fire battle has not been won with man caused catastrophic fire still continuing to increase, even when government land management agencies now know that man not fire is the real cause of catastrophic fire in light fire ecosystems. The science has been done to a large degree since my father's time and I believe that the focus now should be to apply what we already know. We can't expect to solve political problems with science alone, because political problems often require political solutions such as strong citizen activism. I would apply this to Tall Timbers as well. ☺

Since the 1980s, new leadership of Tall Timbers has reorganized and shifted the focus away from fire research and activism nationally and globally into land conservation, a favorite cause of local plantation owners. This shift has been ongoing since Dad's influence began to wane in the late 1970s and ended in the 1980s and is indicated by a change of the name of the institution from Tall Timbers Research Inc. to Tall Timbers Research and Land Conservancy.

In my opinion, several plantation owners saw an opportunity to takeover Tall Timbers as the founding board died out and became weak and to fundamentally change its vision to that of land trust. Fire, plant and wildlife research is still being done of course at Tall Timbers, but in my opinion Tall Timbers has been fast losing its cutting edge leadership role in the field of fire ecology and fire activism, even as the fire battles have shifted to the Western United States and Australia. One old timer told me that in his opinion Tall Timbers is simply repeating most of the research that has already been done.

I of course, would like to see an institutional course correction back to the founder's original intent, more global fire research and fire activism, but I don't think there is much hope of that. It's going to be up to other groups and institutions to pick up the fire activism and public relations ball and run with it now. With the new plantation board members fully in control, I think that they will continue to downplay fire research and activism in favor of even more land trust activity.

Those that are interested in how this evolution of Tall Timbers as an institution progressed from being a cutting edge leader in national and global fire research and activism, to becoming a leader in local land conservation have only to study Crawford & Brueckheimer's The Legacy of a Red Hills Hunting Plantation.

It becomes very evident that the focus of the institution under new management has shifted away from fire research and activism as chronicled after Chapter 5 in this history of Tall Timbers. It is also evident that Tall Timbers influence and leadership in global fire research and activism has declined drastically with the focus on local land conservation to protect large tracts of plantation land.

I am not that concerned about the reduced emphasis on national and global fire research at Tall Timbers because a brief review of the Internet shows that researchers all over the world have picked up the ball that Tall Timbers dropped. My concern is the need for a huge public relations operation to promote the application of that research on a national and global basis to end catastrophic wildfire in light fire ecosystems once and for all.

The current out of control situation in the West and Australia is outrageous and intolerable and in my opinion Tall

Timbers changed course at a time they should have carried the fight beyond the Southeastern United States. Just because the pressure of fire exclusion is off the backs of the plantation owners, is no reason not to pursue the fire battle to West and around the globe until the battle is won.

I have to put some of this responsibility for the shift of focus on Ed and the rest of the founder's board because as Ed's mental and physical health began to deteriorate with old age, he became distracted from the battle against fire exclusion. He and the board also made the mistake of putting local plantation people on the board replacing the founders when they died for funding purposes and not for a vision true to the founder's intent.

What they did not understand was that these plantation owners had a very different vision than the founders. Yes, the founders did have an interest in land conservation to some degree, but I do not think they intended Tall Timbers to become fundamentally a land trust conglomerate either. In my opinion because of the lack of good board leadership, fire ecology is taking a back seat, especially fire activism.

Dad was fond of saying that the research is important, but one must not study something to death as many academics tend to do, but to put that knowledge to work in a very practical way. In my opinion the shift to become primarily a land trust, has overridden and submerged the founder's original intent and Ed and Roy's past work in fire ecology research and global activism.

Today it is widely accepted by many in the government land management bureaucracies that fire is an important part of global ecosystems. One has no further to look than into what the U.S. National Wildland Fire Training Programs are teaching about the need for prescribed fire.

I was fortunate to have been invited to sit in on one of the classes of the Prescribed Fire Training Center here in Tallahassee Florida, in 2013. [92] However, even here the fire expert who gave a lecture and slideshow to the class was

151

discouraged and felt that the fight for prescribed fire was losing ground especially in the Western United States. He stressed the importance to the students that as they rose up into the ranks with influence over policy, that it would be an uphill battle for them to get the kinds of resources and funding for fuel reduction and prescribed fire necessary to even put a dent in catastrophic wildfire.

The 1960s, 1970s and 1980s were a time for growing activism and scientific research. This was a really tough time and many battles were fought with the leadership of government agencies. My father got used to being criticized, obstructed and attacked by anti-fire activists embedded in the leadership of these government agencies in the United States mostly in the Western United States. He said before he died that one of the reasons he made headway was because these activists were older and he outlived them and their illusions about fire.

As I just discussed, Ed and the other founders toward the end of their lives also ran into some political problems within the organization as they aged. There was an internal power struggle after the founders Henry Beadel, Herb and finally Roy had died, involving Ed, then Leon. After Roy died Ed with his judgment slipping, was pushed out and later Leon by new people. Leon told me personally that he had been pressured out many years later.

Jack Rudloe and others from the early years were well aware of this power struggle and the effect it had on the direction of Tall Timbers. Jack reminisces:

"Thinking back, Ed and I share a character flaw, we both annoy people. In his old age he got on the wrong side of a several powerful plantation owners as they sought to gain control over Tall Timbers by forcing out the still surviving founders. But without his personality, charisma and knowledge that attracted people from around the world to his fire ecology conferences and his experiments in land management, Tall Timbers would not be the wonderful and

152

scientifically valuable organization that it is, and the field of fire ecology as advanced as it is today."

Tall Timbers' main building was named after Herb Stoddard rather than Ed, apparently with lobbying by Herb's son Sonny Stoddard, and the old wooden barn converted into a science center was named after Ed. By this time Ed's mental health and physical health due to a car accident was failing so our family went along with this, but the message was clear, and what was done was wrong, mean spirited and creates a false impression as who was central to Tall Timbers in the early days. Ed was also being threatened by at least one of these plantation owners if he did not resign.

The truth is, Ed backed up by Roy, were the central organizing figures in the early years of Tall Timbers not Herb, and having Herb's name on the building gives both the public and employees the mistaken impression that Herb was the central organizing figure of Tall Timbers. I certainly do not want to diminish Herb Stoddard's huge lifelong contributions to ecology and specifically fire ecology, and in this case knowing Herb, I am sure he would have agreed with me on this. Herb had already died previous to this. Herb like Dad always put the truth first in their lives and the same was for Roy Komarek as well.

I found out that Dad really had been affected more than he let on by all his political battles when he broke down as he accepted his honorary PHD at FSU for his life time fire research. He had conditioned himself to have a thick skin toward criticism from friends and foe alike, but was not used to praise and recognition. After so many years of fighting powerful individuals and centralized bureaucracy's ineptitude and incompetence, praise and recognition was a long time coming.

Robert Crawford and William Brueckheimer's book, The Legacy of a Red Hills Hunting Plantation – Tall Timbers Research Station & Land Conservancy, did a good job of setting the record straight. But using a term Dad used to like, it really "sticks in my craw" about the naming of the main administration building and putting Ed in the barn symbolizing Ed as a peripheral figure in Tall Timbers history.

For me this is not about ego, this is about the importance of accurate institutional memory and its effect on the present and the future actions of an organization. Suppressing Ed and Roy's global contributions to fire research and activism may have helped in the shift to a land trust, but it also hurt Tall Timbers as leader in global fire research and activism.

I wrote an op-ed piece published in the Tallahassee Democrat about Tall Timbers and Ed's contributions to Tall Timbers and the world, but I have held back about discussing these internal politics until now, but I want to clear up the historical record on this in print before I die. Here is the op-ed:

How Tall Timbers Came To Be
By Ed Komarek
"Tall Timbers Research Station recently had its 50th anniversary celebration at its headquarters north of Tallahassee and I had been asked to attend to represent the Komarek family in the celebration. Both the Governors from Florida and Georgia were slated to attend, but only Governor Perdue showed up. As I set in the warm sun of a beautiful winter day listening to Georgia's Governor praise Tall Timbers, I could not help but reminisce as to how Tall Timbers came into existence and had developed to the point it had today. My father Dr. Ed Komarek would have been proud.

It had been my father Dr. Ed Komarek Sr. that was the central organizing figure in the creation and development of Tall Timbers for its first 20-30 years of operation. As a boy and young man I was fortunate to have had a front row to the events that lead to the creation of Tall Timbers under the watchful eye and strong hand of my very charismatic father.

My father was fond of calling himself and his friends and associates mavericks, an old cowboy term for a cow or bull that did not run with the herd. It was no small feat for him to

154

organize, backed up by his brother Roy, the very independent and contentious founders to Tall Timbers. My father had his own very popular local TV show called Rural Report with WCTV in Thomasville Georgia that he used to make ecology and weather understandable to local farmers and fishermen. He also used this program to organize and pull these mavericks together under the Tall Timbers umbrella.

Now only was I fortunate to have a front row seat as to the envisioning, founding and early development of Tall Timbers I was even more fortunate to have been in on the early development of the newly emerging field of ecology. Herb Stoddard and Aldo are considered to be the founders of ecology, a new field that emerged out of the old natural sciences, which included many early museum collectors and taxonomists. Herb Stoddard was my father's mentor and treated him like a son and I like his grandson. Herb's Sherwood Plantation and our Birdsong Plantation adjoined each other and as a young boy I beat a trail through the woods to Herb's long house where he taught me natural history.

Old museum collectors like Herb Stoddard, Ed Komarek and my uncle Roy Komarek were quick to grasp the fundamentals of ecology because they were in close touch with nature, while the taxonomists who mostly resided in museum basements were not able to make the jump. Ecology is involved in the study of relationships between plants and animals as well as the plants and animals themselves. Dad used to tell me of the arguments he used to have with his taxonomist friends, who would exclaim. "Show us a relationship and then we will believe in ecology, there is no such thing as a relationship!"

After the anniversary celebration I was given a short tour of the renovated and expanded central Tall Timbers building. It was striking and a bit disturbing to me as I walked the halls that there was really very little there that related to my father, and what was, portrayed him in the background. This was indeed strange because I knew my father to have been the central figure, the man that pulled together the vision, organized the founders and made that vision reality through

155

the wise use of his organizational abilities combined with a very powerful and charismatic personality.

As I left the building I drove home reflecting that institutions just like people forget and distort their past history as they age. Over time institutions lose their founders who either die or are pushed out to make room for newcomers who do not have the memories of how things really came to be. The second generation of institutional leaders without the early memories become vulnerable to internal politics, as special interests inside and outside the institution pressure for the rewriting of history.

I take comfort from the fact that institutions at some point do turn back to rediscover their lost roots and in so doing find a reality check as historians wade through the archives of the institution to sort out fact from fiction. History will show that Dr. Ed Komarek was the one individual that was indispensable to the envisioning, the creation, and the early development of Tall Timbers. My father was not just a founder on the sidelines quietly residing in the background. He was the heart and soul of the institution in its early years."

The larger issue being raised here is the issue of Tall Timbers' focus and direction under the plantation leadership of two or three very influential and wealthy people. Tall Timbers is a non-profit that should in my mind be accountable to the people, its members and other interested parties of which I am one. ☺ Fortunately this book is not published or sanctioned by Tall Timbers, so I am free to say a few things that others can't do.

Now don't get me wrong, I am very supportive of Tall Timbers in general, but I also don't want the institution to become risk adverse as it ages, playing it safe, resting on its laurels so to speak. Strong external citizen oversight is important to the good health of institutions.

I also must express some concern over issues involving some unintended consequences, or maybe even intended ones, involving land conservation easements to protect land in the Red Hills, not addressed in the history of Tall Timbers. The founders of Tall Timbers and this Nation had a very strong sense of private property and the need for good government oversight as I still do to this day.

Decades ago I founded an activist group locally in South Georgia called the Citizens Oversight Group. The group's activities centered about private property issues and civil liberty issues. The founding fathers believed that private property was an extension of an individual liberty, as do I and my local friends.

I have run into problems when individuals representing Tall Timbers came into Grady County and lobbied the County Commission for more land and property restrictions on the citizens of the county, centered about the issue of land use and zoning. Hearings were held around the county and zoning was defeated with about nine to one against zoning.

Zoning is an important part of the UN Agenda 21 initiative, a powerful top down mechanism of social control. The International Chamber of Commerce through its national and local branches drives this New World Order agenda into local government, often very deceptively and without citizen oversight. Economic development and environmental protections are often used as ruses for more social control over people far and beyond what is absolutely necessary.

The other beef I have with this plantation control over the Tall Timbers board of directors is the use of environmental restrictions on private property to indirectly shift the local property tax burden off the wealthy landowners on to small landowners and other property owners. There is a very definite conflict of interest here in my opinion.

Taxes have to be raised and paid by somebody, and small property owners like myself, don't want to give up their property rights, or have to put up with hassles of getting involved in these Tall Timber's sponsored environmental land use programs that are perceived to favor the wealthy. My feeling is that conservation is a good thing, but that if small landowners and property owners are going to have to pay indirectly for land conservation that favors the wealthy, they should be consulted and be part of the process.

I am willing to bet, knowing the founders, that they would want Tall Timbers to have a special fire activist branch or division, in the Western United States and maybe even in Australia, or at least Dad would have had if he had not run out of time. Such a branch or branches would lobby the public, press and government to replace the fire suppression culture once and for all with a pro-fire management culture. The idea is to finally put an end to or at least reduce the destruction of light fire ecosystems in the West and in Australia.

Since the early 1980's I have pretty much been out of the knowledge loop in the struggle to put fire back in its rightful place in earth's ecosystems until I became involved again in writing this book. It's obvious to me because of increasing global catastrophic fire, that the pro-fire battle has yet to be won and is even begun to backslide in some areas. The public just can't let this battle be lost for the sake of nature and humanity and neither can local plantation owners! ☺

It troubles me today that experienced controlled burners or fire managers in government land management agencies are having trouble moving to the highest levels of the these bureaucracies because of fire politics and public ignorance. Even when these bureaucracies try to control burn, the fires get out of control (like happened in Yellowstone National Park and Los Alamos), leveling the places to the ground with catastrophic fire.

Excuses are always the best defense of the incompetent! One of the first laws of control burning is to experiment and make mistakes on a small scale in fire safe environments before ramping up to mess around with hundreds of thousands of acres. If you are not ready or experienced enough, you must put misplaced hubris aside and bring in fire managers with a lifetime of experience and a safe proven track record of keeping fires under control even on very large acreages.

Another problem is central planning and control in fire management bureaucracies. In the USSR it was proven that large scale economic central planning was a disaster. People were moved from the land where they had lived and farmed for generations and put into farming collectives and the same happened with industry and this resulted in the dissolution of the USSR. Central planning seems to work for military style operations for which fire suppression is similar, but central planning is turning out to be a disaster for fire and land management on public lands.

I think these huge government bureaucracies have to be decentralized into locally controlled franchises like we see in the business community or even in the plantation community. After a lot of thought while writing this book, I will share some of my ideas in the next chapter as to how this could be done to integrate a little "folk wisdom" into institutional fire management policies.

CHAPTER FIVE

FIRE MANAGEMENT FOR THE 21 CENTURY

Deconstructing the Culture of Fire Suppression

Over a century of national and global fire exclusion and suppression has led to what can only be called a global catastrophic fire emergency. Governments in countries including the United States and Australia are now under siege, by man caused catastrophic fire. They should declare national states of emergency in order to cut through the red tape, bureaucratic inertia, special interest control, public ignorance and political stalemate to get light fire back into light fire ecosystems.

Over the past 120 years the growing culture of fire exclusion and suppression has evolved unwieldy, dysfunctional government land management bureaucracies unresponsive to reform. To make matters worse, a fire suppression industrial complex worth billions of dollars has developed dependent on the suppression of catastrophic fire. Any and all attempts to move quickly from fire suppression to sound ecological fire management policy have either failed or moved forward at glacial rates because of the above factors. The proof is that catastrophic fires are still increasing in intensity and surface area covered. If real progress was being made overall, the acres being devastated by catastrophic fire would be decreasing rather than increasing.

Any serious attempts at reform are going to have to have broad public support and that has been a long time coming in spite of the damage to life, property and nature's ecosystems. It's sad but true, that the pain is not yet great enough to overcome the obstacles to good fire management. Good fire management will come; it's only a matter of time before the suffering from catastrophic fire becomes severe enough to force change and undo over 100 years of fire suppression propaganda. The question really becomes is there anything we can do now to facilitate reform and reduce the damage to both man and nature by removing obstacles to good fire management?

The book Burning Questions by David Carle has a section on my father and this one case that well illustrates that it takes much more than science to change the fire suppression culture in large government agencies. It's going to take powerful activism by both individuals and institutions following in the footsteps of people like Ed Komarek Sr. The fact that catastrophic wildfire is on the increase in much of the world is an indication that now that we have the science done, emphasis should be shifted to powerful determined activist solutions to change the culture of fire suppression to that of one of a culture of good fire management.

"In December Komarek invited Weaver to speak at the 1963 Fire Ecology Conference. "I have been much impressed, both by your writings, as well as the practical application of fire on Apache, Klamath and Colville Reservations. The many objections raised by some foresters to the use of fire in Ponderosa seem awfully reminiscent, even to the actual phrases used, to what Mr. Stoddard and I have had to put up with, until recently. Now one would think the Forest Service invented the use of fire in Longleaf and Loblolly." Komarek mentioned he had hoped to see Dr. Biswell during his western tour, but found out when he got to Berkeley that he was in Greece.

Weaver requested to BIA permission to attend the conference. His letter to his supervisor revealed concern that the request would be denied; Weaver made it a personal plea: "I have firmly in mind your letter to me of August 31, 1962, and my reply of October 5. In it I called your attention to the fact that my past advocacy of more research on fire in ponderosa pine has frequently made me the center of attention with respect to this subject." He was aware of the "austerity program" within the BIA at the time, but "I would be anything but frank if I did not indicate that I would very much like to go to the conference. I have always wanted to visit the Southern Pine region and this looks like the best chance, if not the only chance, that I may ever get, for I am nearing the end of my career as a forester. I will not harm the Bureau's relationship with anything that I may present. In fact, I may do them some good."

Weaver, on the next day, also sent Komarek word that Biswell was back from Greece. Komarek, assuming that Weaver would attend, told him, "I am now hoping that we can get Dr. Biswell to also discuss fire and (southern) California at the conference. I heard some vague references among forest service personnel that he had some unhappy experiences because of his views, probably similar to what Mr. Stoddard and I had to contend (with) in years gone by."

Weaver was denied permission to attend. He sent that disappointing news to Komarek on January 30, 1963: "The official letter refusing permission showed quite plainly that the Office does not want me to discuss the ecology of fire in ponderosa pine under any circumstances. After enumerating the various jobs that I am expected to do this spring, the letter suggests quite pointedly that there will not be time available for me to go to Tallahassee."

Komarek had to wait a week to let his anger cool before he wrote back. "We are mighty concerned about you not being allowed to attend, even at our expense and your time. Frankly, if you were not so near retirement I would force the issue" by seeking support from friends who were senators. "I had been somewhat afraid of this, "Komarek continued. "Mr. Stoddard and I had hoped that this sort of thing was behind us. You see some 25 years ago I was even threatened with arrest for burning a client's land with his express permission. I would have written you earlier but I am still a bit hot under the collar. Mr. Stoddard and I have leaned over backward to be more than fair with the various services. However if some of them continue to try un-American tactics we can sure have a good discussion in Congress when appropriation bills come up for hearings."

Harold Weaver was allowed to attend the Third Tall Timbers Fire Ecology conference in 1964, where he spoke on "Fire and Management Problems in Ponderosa Pine."

David Carle gives some information that shows what a powerful activist person can do in their own right if they are fearless, dedicated and have the passion to use force when necessary to make beneficial changes in human culture and consciousness. False beliefs once firmly entrenched die hard! The struggle today to end catastrophic fire in light fire ecosystems, is no easier today than in my Father's time, the only difference is that the battle lines have shifted somewhat and the obstacles in some instances have become more subtle and devious.

"More than anyone else, E.V. Komarek . .. promoted the concept of fire as one of nature's most potent evolutionary and ecological forces" Komarek directed the Tall Timbers Research Station for twenty-one years. He ultimately delivered lectures in twenty-four states and fourteen countries and was awarded an honorary Doctor of Science degree from Florida State University. Komarek's papers were donated to Tall Timbers in 1987 and became the genesis of a fire ecology database, named for him, that holds over 12,000 records and can be searched on the Internet. I believe Ed Komarek could sell a forest fire to Smokey Bear," James Stevenson said at the 1989 Tall Timbers Fire Ecology Conference that honored Komarek."

Obstacles to Good Fire Management

In writing this book, I can see how far scientific research of fire in nature has come since I was a bystander in the battle to put fire back into the environment. Anybody that does some research on the Internet or works their way through the Tall Timbers' Fire Conference proceedings can see the huge volume of research on light and catastrophic fire to date, either in the scientific papers themselves, or to the bibliographies on which the articles are based. But more research is not going to solve the problem. It's also going to take political solutions to solve some of the political issues involved. It's going to take strong grass roots educational and political activism to get things moving in the right direction. Here are some of the obstacles involved.

1. Need for continued good scientific research on fire. 2. Lack of awareness of native peoples contribution to good fire management. 3. Lack of integration of native knowledge and wisdom into fire policies. 4. Bureaucratic inertia. 5. Academic and bureaucratic capture. 6. Lack of consistency of fire management. 7. Excessive central planning and control. 8. Lack of public awareness and trust. 9. Risk

After the catastrophic fire season in Arizona and in California in 2013, because of the severe loss of life and destruction of property, people as in other years speak out about the need for reform, but it seems the reform is only incremental and interest soon wanes. In this USA Today article titled Wildfire experts call for more controlled burns, some of these issues are addressed. [93]

"PHOENIX -- As families begin to return to the fire-ravaged communities of Arizona's Yarnell and Peeples Valley, and as investigators delve into why the blaze killed more firefighters than any U.S. wildfire in 80 years, fire experts are renewing calls to make prescribed burns easier to accomplish.

Days after the fire, President Barack Obama said the incident "will force government leaders to answer broader questions about how they handle increasingly destructive and deadly wildfires."   Firefighters say they know the answers: Ease environmental restrictions and spend more to clear brush and light prescribed burns. They are calling for political courage."

"The reality is ... there are not enough money or people or resources to go around to make a dent," Broyles said. Regulations to protect habitats and clean air often interfere, as regulators balance protecting endangered species and people's lungs with the need to thin forests. It can take up to four years to get approval to do a burn, which has to happen in a three-month window in the fall. Even then, crews can burn only a few hundred acres at a time, and the backlog of acreage is in the millions. One in five burns is halted midway by state regulators, Hughes said."

"Hughes and others suggest these reforms for preventing future tragedies like the Yarnell Hill Fire: • Loosen restrictions in the Clean Air Act to exempt prescribed burns, or at a minimum make them easier. • Loosen restrictions protecting endangered species in the National Environmental Act, particularly when their native habitats rely on periodic fires.• Direct environmental regulators to follow policies uniformly.• Increase federal and state budgets for fire prevention.• Convince environmental activists that improving forest health is better, not worse, for the environment, and convince the public in fire-threatened areas that controlled burns are vital.• Pass and enforce stricter laws requiring property owners to clear dangerous fuels on their land and create "defensive space" from structures.• Hire more "hotshots" or specially-trained crews to do brush clearance and prescribed burns.• Promote volunteer crews to clear fuel."

The Need for Good Scientific Research on Fire

I think we can safely say we are well on our way to having removed the first of seven major obstacles to proper global fire management, with evidence being the first obstacle to overcome. It was a long hard fight by the pro-fire advocates of my father's day to get us this far. The way it looks to me, is that we now have plenty of good scientific research and evidence proving the importance of fire in nature and the need for the use of prescribed fire in most of the fire environments on earth.

Public Recognition of the Contribution to Fire Ecosystems by Native Peoples

The second obstacle just now being overcome is a deep profound appreciation of and for the intelligence of indigenous peoples among research scientists and land management organizations. It is just beginning to be recognized that native peoples have a very deep and abiding awareness of the land upon which they live that developed out of necessity, but it has yet to be very well incorporated into institutional public awareness. The fire skills these peoples developed and practiced to create a great diversity of habitats and ecosystems over tens of thousands, if not hundreds of thousands of years are still not well understood by modern scientists. Nor are they being integrated very well into public land management policies and procedures.

Fire Ecologists even today are having a tough time trying to figure out where nature's fire influence on global ecosystems ends and early man's influence begins. Even today some scientists are questioning what should be obvious, that man has significantly altered natural ecosystems for a very long time. They suggest that lightning started most fires in the past history of man and are having to be proved wrong by research using lightning fire ignition statistics across the globe. So we see this second obstacle still has some legs. ☺

Putting Fire Managers Back on the Land They Manage

The third obstacle to be overcome is how we integrate what we are learning about native people's intimate association and knowledge of the land into modern fire management institutional policy. One of the major factors is native people's exquisite detailed knowledge of plants and animals, their habitats and fire ecology. The children learn from their parents by example through a very ancient apprentice system, where knowledge is passed down from generation to generation.

It is widely understood that children gain very basic fundamental knowledge, understanding and skills much more quickly than adults from their surroundings. As people age they gain in experience, but most lose perceptual abilities and sensitivity to their surroundings due to increased thought activity or mental dialogue. How can we even begin to compare the fire skills of a well-educated academic raised in the city with several hundred hours of fire training, with those of a primitive farmer or hunter-gatherer in intimate continuous contact with his or her land, fire and the changes to the land after fire for a lifetime?

Restructuring Public Land Management Bureaucracies

The forth obstacle to be overcome is a problem common to organizations and institutions in general and that a systemic failure to adapt to change. It's a problem that the natural world takes care of through a kind of adapt or die policy, that is true also of the free market system of the business world that attempts to mimic nature and natural human tendencies in a way. With some regulations to keep predatory capitalism in check, the system works and much better that a central planning economy.

The problem with government institutions is that as they grow they tend to escape public accountability and increasingly are more interested in self-preservation of jobs and positions than serving the public. Another problem is that government institutions tend to become less adaptable to change as they age and fixed in their thinking, just as happens to individuals. In human society as in nature, this problem is solved as old people

167

die to be replaced by their children. But institutions and corporations in theory live until they become so out of touch with reality that some catastrophe like war takes them down, or forces them to restructure.

Large government land management agencies are no exception to this rule. The fact that huge fuel loads are continuing to build on the public lands resulting in more and more catastrophic fire caused by fire suppression is still not enough to rapidly shift away from fire suppression to extensive prescribed fire. In a sense they have gone to war against fire and as fire damage increases, they will either lose the public's confidence and support, or they will be forced to restructure by the severe public and environmental consequences.

These bureaucracies have become trapped by the Smokey the Bear propaganda monster they themselves created in the public mind. The destructive meme that all forest fires must be suppressed, combined with the fire suppression industry now dependent on fire suppression income, make needed change difficult and almost impossible.

The struggling fledgling prescribed fire divisions within these organizations have little political clout to obtain more funding in the face of a powerful fire suppression lobby also under financial pressures because of tough economic times. In addition, centralized control and central planning work well for military operations and for firefighting, but poorly for extensive controlled burning programs needed that seem to require more decentralization, intimacy with nature and knowledgeable community involvement.

Harold H. Biswell, in his book Prescribed Burning in California Wildlands Vegetation Management, lists as his last chapter excuses and obstacles that incompetent bureaucrats and bureaucracies can come up with as not to control burn and then proceeds to demolish the excuses. ☺ I suppose the list is really endless, but Biswell identifies the major excuses of the incompetent, impotent and powerless to do anything bureaucrat!

"1. The Idea That All Fires Are Bad. 2. Confusing Prescribed Fires With Wildfires. 3. Too Much Danger of Fires Escaping Control. 4. Too Much responsibility. 5. Dislike of Smoke From Prescribed Burning. 6. The Public Won't Let Us Burn. 7. We Need More Research. 8. There Aren't Enough Burn Days. 9. Prescribed Burning Is Too Costly. 10. We Can Lose Our Jobs. 11. There Is No Money For Prescribed Burning. 12. Negative Influence of Powerful People. 13. Let It Be an Act of God."

The Issue of Academic and Bureaucratic Capture

The fifth obstacle we need to overcome is the problem of academic and bureaucratic capture in which we study the problem to death and do not apply what is learned to remedy the situation. There seems to be no lack of detailed scientific papers today as to the need to move from fire suppression to fire management. However, what good is this knowledge if it does not lead to movement away from fire suppression to fire management and prescribed fire?

Science does not tell us how to act, but culture does. Science can have some effect on culture as long as the political aspects of culture are not so strong as to deny good science. In a case where the political aspects trump the science, we have to deal with the political aspects as we deal with the science. If good fire management and good public fire policy was based on science, we would be much further along than we are today. Fire historian Steven Pyne accurately states in his article Fire, science and culture:

> "The fundamental questions lie not in science but in politics – that's why the national investment is poor. The science gets funded best when it adorns political purposes. The field requirements lie in a learned sense of how the fire guild operates and why. The chi-square conclusion is clear: while science can counsel, it cannot choose. "The science" doesn't tell us how to act. Culture does." [94]

"Prescribed fire resides overwhelmingly in the southeast. It continues for traditional reasons, not out of scientific discovery. People have always burned here. But old habits of woods burning can't continue any more than open-range ranching. What modern research has done is sharpen the prescriptions and help reconcile an inherited practice to a fast-changing environment. It has improved field operations indirectly by boosting technology. And in a historic reversal, it has sanctioned practice where it previously condemned it."

Long Term Consistency of Fire Management

The sixth obstacle to sound micro and macro fire management is the problem of long term consistency of fire management. In early hunter-gather and farming, society's knowledge of land management is passed down from generation to generation through the family. The children learn from their parents the history of the land and so come to understand what their contribution to that history is going to be.

The situation is different with institutions where people come and go and take their institutional knowledge and wisdom with them when they depart leaving a gap in institutional historical knowledge. History is important because it's how we orientate ourselves and find our true calling in the timeline of history. Land management is a form of artistic expression where each land manager paints upon the painting that came before, be it a poor painting or a masterpiece.

It's extremely frustrating to me to see a very good land manager using fire to create a masterpiece, a model of diversity of plant and animal species in a micro or macro ecosystem, only to have that masterpiece neglected and painted over by a land manager that does not even recognize the masterpiece and elements that he is painting over. Such a poor land manager in his or her ignorance thinks what they are doing is just as good as what came before.

One of the things that helped me a lot in understanding the history of our Birdsong Plantation was the Dickey daily diaries written during the Civil War period. They were short entries, but they recorded a consistency of land and agricultural management during the time period valuable to the history of the time and future generations. I combined that thought with a ship's log and came up with the idea of institutional land management log for the parcel of property being managed.

No matter if it is a quail plantation or Yellowstone National Park, if a detailed log exists of the land and fire management year to year, including pictures, this would go a long way toward maintaining consistency of fire management. When a new manager arrives to manage the parcel of land he studies the log so as to build on this consistency of management and not destroy prior works of others that came before out of ignorance, but only upon considerable deliberation should a change in management take place.

Additionally, a published log would allow better oversight of management by supervisors and the public. We could go even further and do something like Google does by traveling down roads and through neighborhoods to build a database of the situation on the ground. This could be done every year or every several years and give a good record of the management of a property.

If someone has spent a lifetime creating a masterpiece of art, it's pretty stupid for a second class inexperienced artist to come in and paint it over without even recognizing or appreciating the masterpiece being painted over. In the old world, apprentices would spend years copying the paintings and techniques of the masters, before going out to try to paint a masterpiece of their own. In the case of land management, a person gets a little academic training and field experience and they think that they are already a master, or as least soon to be one. ☺ Sorry, but that's not good land management.

Rebuilding Public Awareness and Trust

The solution to the eighth obstacle (that of public ignorance) is to rebuild public awareness as to the importance and the role of fire in nature based on good scientific investigation and research. It's going to be a public relations offensive the equivalent of the propaganda operation that government agencies like the Forest Service mounted and implemented in the 20th century against fire and the public consciousness.

For heaven's sake, some of our own environmental groups are working against putting light fire back into light fire ecosystems because of this public ignorance. For example in the USA Today article called Wildfire experts call for more controlled burns, it shows a good chunk of the endangered Spotted Owl habitat was destroyed because of public and environmental group opposition to prescribed fire. [95]

> "A 1996 blaze that destroyed thousands of acres in Four Peaks shows what can go wrong without a burn. The year before, a controlled-burn permit was denied to protect the spotted owl. The entire range went up, and with it, the owl's habitat. Fires in Arizona at Alpine, Mount Baldy and Prescott show that brush mulching and prescribed burns stopped worse devastation."

Key to this is that the agencies involved must admit and not paper over their responsibility for the assault on public consciousness and the suppression of scientific research into fire. Just as happens with individuals, organizations and institutions must take responsibility and admit error if they are to reform themselves and regain public confidence. Part of this reform process is to make sure that in institutional publications, as well as interactions with the press, that an accurate account of history be recorded for the present and for prosperity. I see a considerable amount of denial and other attempts to downplay systemic failures of these institutions in the past that even shows up on Internet sites such as Wikipedia.

The first step toward rebuilding public awareness and confidence must be the unequivocal admission, confession and regret by the institutions responsible for the damage to public awareness by 120+ years of propaganda assault. The U.S. Forest Service, U.S. Park Service, Fish and Wildlife Service and Bureau of Land Management must research the fire history in detail of their organizations and not just expect independent organizations like Tall Timbers to do it for them.

Nobody likes reviewing past misdeeds or having to confess to the world the damage they have done and why. Just as with individuals, there is no other way for institutions to begin and implement rapid serious reform in the present if the past continues to distort the present and future.

It's going to take a tough hard hitting approach to prescribed fire activism in the mainstream and alternate media to reverse this 120 year old assault on public consciousness. I figure a good example of the kind of approach needed would be my own press release for this book. Such a press release might go something like this:

For Immediate Release
Prescribed Fire Activist Blames Smokey the Bear for Catastrophic Wildfire

Fire activist Ed Komarek has just released his book Fire in Nature, A Fire Activists Guide that blames government misguided fire exclusion policies and propaganda the past 120 years for horrific wildfires in the Western United States and Australia.

Ed Komarek says, "It is not fire, arson, global warming drought etc. that is primarily responsible for catastrophic wildfire. He and other fire ecologists claim that the real cause for most catastrophic wildfire is the growing unnatural fuel accumulations caused by fire exclusion in the Nation's forests and grasslands that lead to catastrophic wildfire."

According to the statistics in Wildfire Today [4] the average number of acres devastated by wildfire in the United States lower 48, has risen steadily from above 2 million acres in 1990 to above 6 million acres in 2013. An article in Headwaters Economics [5] states that U.S. National wildfire fighting costs have averaged $1.8 billion annually for the past five years, with costs are set to explode to between $2.3 and $4.3 billion.

We are told that the science has been done, the jury is in, and the verdict is that light fire is essential to healthy Ponderosa Pine forest in the West, and Longleaf Pine forest in the East etc. These are fire adapted ecosystems that need fire to survive. Light fires sweep away the dead yearly debris accumulation without hurting the trees, thus maintaining healthy species diverse, ecosystems.

Ed conclusively shows in his book that fire has sculpted ecosystems for hundreds of thousands of years, helped along by evolving mankind the past one hundred thousand years at least. With this book, Ed follows up on his father's earlier fire ecology work throwing his hat into the ring to join in the still ongoing fight to put light prescribed fire back into our remaining fragmented ecosystems worldwide.

Ed's book is for sale on Amazon both in print on Kindle, but free as a public service to all on the book's website. fireinnature.weebly.com/

Risk

David Carle points gives us an example of how as old obstacles to prescribed fire diminish, others rise to the fore continuing to hinder change. This 9th obstacle of risk and short sighted cultural behavior by the public, bureaucrats and politicians is well stated.

"Knowledge reduces risk, yet risk remains one of the major obstacles slowing the implementation of the National Fire Plan. After forty years with the U.S. Forest Service, Bob Mutch retired in 1994. Now a private fire management consultant (his work, in 2001, took him to Italy, Ethiopia, and

Mongolia), Mutch is particularly concerned about a double standard that impairs "our ability to prescribe fire on the landscape on a large enough scale to sustain healthy systems. A prescribed fire can be well-planned and well-executed by qualified people, but the moment something starts to go awry the support from politicians and the public is quickly lost.

In contrast, practically any professional strategy can be adopted in suppressing a wildfire and vast amounts of money can be spent in implementing that strategy. No matter how adverse the outcome … politicians and the public generally side with the firefighter. For example, in a Malibu neighborhood in 1993 where practically every house burned to the ground, the signs on the street said, "Thank you, firefighters." This double standard is part of our tradition and culture."

Now that we have defined these obstacles to sound global fire management, let us try to continue to expand our understanding of each to these to find a better way forward to remove the obstacles much quicker than they are presently being removed. Time is not on our side as more and more damage to the environment is accumulating through catastrophic fire and lack of detailed and intimate knowledge and policy for plant and wildlife management in fire ecosystems. Every day that we drag our heels in confronting these problems results in more damage to both ecosystems and human society.

Let's get back to this issue of bureaucratic inertia and lack of political will. We would no longer be having these serious conflagrations in the western United States if the land management institutions and political institutions were not still seriously dysfunctional and unhealthy. When an individual is mentally ill and dysfunctional, he goes to a mental health professional, and I suggest that some kind of equivalent of a mental health professional be consulted by the institutions in question as to how best to achieve wellness. I suppose one way can be to bring in independent consultants to give seminars on institutional health, who can supervise and grade the institutions and publish the results. ☺

On the issue of incorporating folk wisdom into modern ecological management, the Southern Quail Plantation can offer public institutions a better way forward. On well managed quail plantations in Southeast Georgia, the plantations hire independent foresters on a fee basis to manage the timber and to burn the woods where they are not able to do so themselves competently. In fact, the better managed plantation offers a template for better healthy institutional management of land that incorporates intimate native knowledge combined with modern professional advice and support. One way to heal an unhealthy institution is for the unhealthy institution to search out, study and emulate healthy institutions.

The Art of Managing Longleaf by Leon Neel should be required reading for public institutional land managers and their supervisors. The Stoddard-Neel Approach to timber harvesting and wildlife management simulates natural process to bring in income from timber, while at the same time improving forest health, density and quality of trees and improvements in plant and wildlife habitat. This approach requires great awareness that includes examining each individual tree on the property to be thinned when appropriate or if healthy and straight be allowed to grow so as to improve the whole forest. I quote from the Art of Managing Longleaf as to forest and land management on Greenwood:

"But neither of us was quite prepared to make to make sense of the unusual beauty of Greenwood. Yet that aesthetic reaction is the foundation of the Stoddard-Neel Approach. Many foresters are quick to dismiss aesthetics as a proper measure of good forestry, or they are uncomfortable with a set of values that seems not only far removed from the efficient production of timber, but sometimes even hostile to it.

Leon Neel and Herb Stoddard before him, however, used the look of the woods as a gauge to measure their health. First, he instructed, a healthy longleaf woodland allows one to see a great distance through the trees but also always to see trees. As he makes clear in the memoir that follows, that long look through the forest, which early quail hunters in the region prized, is an important metric of several critical functional aspects of longleaf ecology and management

Second, Leon was quick to point to the many small patches, or "domes" of regeneration that dotted the understory. These are where the future of the forest, he insisted, as important as the gnarled flattops to his practice of forestry beyond one generation. When inspecting the woodlands he manages, Leon is quick to admire good, thick patches of regeneration in the small openings made by his careful forestry practice, and we have come to take joy in them too. Third, was the diversity of the understory as it existed across a landscape gradient defined by altitude and moisture.

We stopped frequently to admire the seasonal blooms of orchids and other wildflowers and to note how the dry uplands gave way to thicker growth in the hardwood drains that ran through the Big Woods. While Leon insists that you can gauge the health of a longleaf woodland by how it looks, he also has taught us that no two healthy longleaf woodlands look exactly alike. Indeed part of the aesthetic joy to be taken from these landscapes comes precisely in recognizing how geology, soils, micro-climates, moisture gradients, and disturbance histories sculpted them into a once-vast mosaic.

Many of Stoddard's and Neel's most important ecological and management insights came from plumbing their sense of the diverse beauty of these woodlands. This is one reason why Leon has consistently insisted that the Stoddard-Neel Approach is an art, not an exact science."

At the deepest levels, land managers are landscape artists painting on the living canvas, just as other artists paint on canvas or shape stone. If the new land manager is not intimately familiar with the land past and present, he might just be painting over a masterpiece, a real national treasure. My father saw this happening when he was collecting mammals in the Great Smokey Mountains in the 1920s, the part that was in the process of becoming a park.

Because my Dad had empathy for the people around him as well as with nature, he really felt for the displaced local people he had become good friends with. These families, who for generations had managed the land in this area, were being thrown off their properties and he deeply felt the suffering this breaking of their deep bonds to the land.

He also watched as I was growing up how park managers lost much of the qualities of the land they so admired because they did not realize that one cannot separate the land manager from the land any more than you can separate an artist from his painting. In my opinion and in Dad's opinion, the Park Service and the public were as much the losers for this as were the local people displaced from the land they had grown to love and appreciate for generations.

This greater awareness and intimacy with nature continues into the management of wildlife habitats, ground cover, cover from predators, accessibility and aesthetics for hunting etc. This is something that is common amongst native peoples. The best way to get this kind of profound understanding of the land is to grow up on the land where one works and to apprentice with a master who may not even be an academic. The first person to let a fire get out on Tall Timbers was a PHD and not a locally educated land manager. ☺

Academics are good at specialization through indoctrination but usually poor at synthesis because academic indoctrination gives academics tunnel vision and an inability to understand and comprehend the big picture into which they are imbedded. I suspect that the Great Depression did my father a great favor by removing him from academia while still an impressionable young man and to cast him to the likes of such a mentor as Herb Stoddard who was a self-taught scientist who greatly expanded Ed's and Roy's horizons at a critical juncture in their lives. Leon Neel got some academic training as a forester, but he already had an intimate knowledge of the land in the Thomasville region before he too was mentored by Herb.

A second step in the institution healing process for public land management institutions should be that once error is openly and widely acknowledged that public education policies be implemented to restore public awareness of the importance of fire to the environment. A part of this public outreach should be to advise public landowners through workshops and through the press on how to protect their property from wildfire conflagrations.

The same Forest Service departments that issue fire permits sometimes are available to the individual private landowner for a visit and free consultation on how to reduce wildfire risk. These consultations could be expanded to include house construction, debris removal, thinning as well as prescribed fire. Folks should be encouraged not to plant high fire species of trees close to the home. These trees are adapted to burn out the competition, in this case the home. ☺

Partnering with insurance companies that have a lot to lose from wildfire should also take priority. Much property damage from wildfire conflagrations has to do with flying embers traveling ahead of the fire. When these burning embers land on debris and dry vegetation near the home, shingle roofs, or in the private forest around the home, they ignite and burn the land and home. The fire itself may not even reach the home. The work of doing an assessment and debris removal can be shared with other partners who have a common interest to protect their investments.

This same partnering with insurance companies can result in incentives to the landowner to maintain a fire-safe environment through reducing premiums for those that keep a fire-safe environment and increasing premiums for those that don't. Such creative measures such as these will help increase public awareness and action having the additional effect of improving relations with the large public land management institutions in the area.

This is the sort of creative thinking forced upon a healthy individual or institution by necessity when there is not a lot of money to throw at a problem. On the other hand, an unhealthy institution or person, tends to make excuses like, we don't have the resources and manpower, just give us more money and resources to fight fire and things will be fine. Or when a prescribed fire gets too hot and kills or damages pine trees, or gets away creating a massive catastrophic fire, an unhealthy institution says, "we were not able to get the burning done in winter, so we had to burn in May when the candles (new growth) were out on the pines. Because of smoke regulations we had to burn in the heat of the day and light the head-fire away from the road." (I am thinking of a controlled burn, I noticed recently on my way down to Saint George's Island with a friend this spring.) ☺

A good indication of a failing management team is the degree to which they make excuses for their mistakes and errors. Often the ingrained incompetence is so severe that in a company the only way to rectify the problem is to have a hired gun infiltrate the company to observe, fire, and reposition employees to return the company to health and competitiveness. The problem with government institutions is that they are not in competition with each other as happens in the free market, so there is not this check on incompetent management that will turn management around before external conditions create catastrophe.

How are you going to be able to regain public confidence if entrenched management is refusing to adapt to change? What good is it going to do to try to educate the public when the institution has lost the public trust? I think the old adage, "Take the timber out of one's own eye before attempting to take the splinter out of the eye of another," applies here. Self-education and understanding go hand in with teaching!

Getting back to public education and activism, there are some bright spots beginning to develop across the country resulting from private property destruction by wildfire. A group known as Fire-wise communities [96] is doing good activist work to get home owners to clean up the land around their houses and make their house more fire resistant.

"Find out what the experts know about the best way to make your home and neighborhood safer from wildfire. From the basics of defensible space and sound landscaping techniques to research on how homes ignite (and what you can do about it), there are tips, tools and teachings you can use!"

I am also reminded that an insurance company is working with the Nature Conservancy to promote public education on the need for fire and how to protect property from wildfire. So in the continuing battle we can find allies with deep pockets to help counter the special interests deep pockets who continue to profit from man caused catastrophic wildfire. I quote: [97]

"Jamestown – A team combining ecological know-how with hotshot firefighting is being deployed in Front Range forests to try to address Colorado's wildfire predicament needing the purge of fire but not wanting it. The team members, assembled by the Nature Conservancy, carry the axes and chain saws used to fight fires. But they also light fires, 47 controlled burns over the past four years."

Of course this is an insignificant amount against the many millions of acres needing this sort of treatment in the Western United States but it's a start I suppose. All this seems to be occurring within the context of great upheavals in land use and land management. Stephen J. Pyne who has done so much to educate the public and bureaucracies though his many books has addressed another complicating factor in land management and fire management in his book Tending Fire.

"The great upheavals of land use continue. Among the Big Four both the re-colonization of rural lands and the decolonization of public lands is accelerating, leaving for the

181

latter, only the rump of imperial institutions. The era of the public lands as a big-government commons overseen by state-sponsored forestry is dying on its feet. This is true in the United States as it is in Australia, Canada and the rest of a hollowed-out European imperium. The means and ends of fire management on those lands reflect these momentous trends. Whether by creep or rupture, from sheer accumulated internal strain or from sharp external stresses, the American way of fire has been significantly altered. Subjected to enough heat and pressure, even granite can melt and warp. The Forest Service that joined a national mobilization to fight the 2000 fires is not the organization that sought to suppress the Wenatchee and Southern California conflagrations of 1970.

Which is to say, the institutions, not merely the politics of fire protection, have rapidly and probably irreversibly undergone a metamorphosis. The evidence lies all around. Privatization, partnerships, the devolution of political decision-making to more local jurisdictions, indigenous land claims, and near civil war over the destiny of the public domain—all are changing the attributes of how governments administer these lands and how they cope with fire. Such reforms have challenged not just the hegemony of the Forest Service but the command-and-control model of federal administration itself. They are not restricted to the quirks of fire management. In their aggregate they promise for public-land fire protection a reformation equivalent to welfare reform."

Creating a Franchise System of Fire Management on Public Lands

It's been a learning process for me to catch up with the learning curve on fire management since my childhood and youth as I write this book. I think what we need to begin to seriously think about is major long term restructuring of land and fire management bureaucracies in order to address these multiple issues and obstacles outlined in this book.

We need to decentralize these large public special interest controlled dysfunctional land management bureaucracies into a franchise organizational system with experienced local control and management. Maybe it is already happening as Stephen Pyne suggests in some kind of organic fashion. The individual franchises would only be controlled on the macro level by implementation of scientifically recognized general land management and fire management standards of quality control like what we have in many types of businesses today. If the franchise is not up to these quality control standards then it is replaced by another franchise that can do the job.

To a degree, the individual franchises will cooperate together and would be in competition with each other with the public being the ultimate arbiter to determine what these standards are. Each franchise would be influenced by both local and national public interest with the overall public interest always at the forefront. Local timber companies, for instance, cannot be allowed to pressure for local timber cutting in excess of what is good sustained forestry and wildlife practices. On the other hand, lobbying by national institutions can't be allowed to micro-manage the franchises either.

Stephen Pyne point out in his book Tending Fire that:

> "Arguably, the momentum for fire management no longer resides with the large bureaus at all, but with private landowner, NGOs, and the like, which stand outside the blood feuds. The comfortable polarization that has dominated discourse for the past few decades and mustered every interested party to one polarity or the other is an anachronism. (So fixated are some critics on hindering misuse that they prevent legitimate use as well. Fearful, for example, that good forest thinning and burning might lead to bad, they would rather ban the practice altogether.) Increasingly, fire management is the work of consortia, of projects conceived, staffed, and funded by public, private, and non-governmental organizations. In this game no one holds a stronger hand than the Nature Conservancy (TNC)

TNC's reach is astonishing. It cultivated fire expertise, originally, because many of its holdings were prairie or fire-adapted savanna and required burning. What astounds, first, is not the volume of burning, however, but its variety—from Kansas prairie to Carolina sandhills to Albany scrubland; and what astonishes, secondly, is the density of the institutional support behind it. TNC trains its own crews, devises its own fire plans, negotiates with neighbors about its fires (and their smoke), contacts for what science it requires—the whole constellation of fire practices. In effect, it has created and NGO that does what, over the past century, only government institutions had increasingly claimed for themselves alone. Perhaps not surprisingly, TNC's fire offices are sited on the grounds of the Tall Timbers Research Station, that vital critic of state-sponsored fire programs."

The TNC's offices have been moved from Tall Timbers. On the Nature Conservancy web site, [98] the article titled Fire and Conservation What We Do explains their advanced concept of integrated fire management.

"In many of the places where we work, fire can be a conservation threat, a natural and even necessary ecological process, and an irreplaceable, life-sustaining tool for rural communities. Where the fire-related needs of ecosystems and people are at odds, The Nature Conservancy has found that it is possible to reconcile these needs through a framework called Integrated Fire Management."

"Integrated Fire Management is defined as an approach to addressing the problems and issues posed by both damaging and beneficial fires within the context of the natural environments and socio-economic systems in which they occur, by evaluating and balancing the relative risks posed by fire with the beneficial or necessary ecological and economic roles that it may play in a given conservation area, landscape or region."

The Nuclear Option – Lawsuits to Force Fuel Reduction and
Proper Light Fire Management

Lawsuits to force public land management agencies to put light fire back into light fire ecosystems could be a very important tool in our pro-fire activist arsenal. Lawsuits could also remove bureaucratic and private obstacles to fire hazard reduction that keep cropping up and secondarily be used as a means for further public education. These lawsuits could focus on two separate but interconnected issues, one being fire hazard reduction and negligence and the other mismanagement of public and private lands endangering light fire ecosystems.

Carnival Cruise Lines has faced a lawsuit for failing to address a known fire hazard risk caused by fuel leaks in its engine room that resulted in a fire on one of its ships. Management had known about the risk of fire and had done nothing about it prior to the fire aboard ship. My thinking is that public land management bureaucracies have known for decades that fire suppression leads to increases in fuel loads in light fire ecosystems causing huge catastrophic fires that destroy lives and property.

Like Carnival Cruise Lines they have been negligent in allowing these fuel loads to accumulate, granted that they are often hamstrung by other private and public agencies like the EPA and even Congress. In my opinion, these other players are also participating and supporting the mismanagement of public and private lands and also could be subject to class action lawsuits to cause them to stop creating obstacles to overall fire hazard reduction.

In the case of the mismanagement of light fire ecosystems, this would appear to be a violation of the public trust to competently manage public lands and maintain biodiversity. Clearly we are dealing with an entrenched system or culture with many different players contributing to the problem, so lawsuits should be both narrowly focused and broad spectrum to clear obstacles to adequate management of light fire ecosystems.

There is plenty of finger pointing among all players, but there is plenty of blame to go around causing these outrageous fuel buildups resulting in catastrophic fire in light fire ecosystems. The end result is that as shown by the continued increase of catastrophic fire globally, that the current situation needs a series of powerful shocks to bring about major changes in fire management policies. A series of high profile lawsuits could make all players supporting dysfunction on notice, creating a willingness to support change on all fronts.

The Need for a Pro-fire Global Activist Organization

In doing the research for this book, I have not found a pro-fire global activist organization completely devoted and focused to putting light fire back into light fire ecosystems to stop the ongoing catastrophic fires around the world. What I see is individuals and groups spread out here and there, raising their voices independently in the press on the importance of fire in nature and the need to end these catastrophic fires through activism and political action. I think what is needed now is an independent organization completely devoted to ending these man caused catastrophic fires. So far the Nature Conservancy's global efforts in regard to fire are the closest to what I have in mind, but will this be sufficient?

The Nature Conservancy has a broad focus on land preservation as a whole, but on the other hand, their efforts are really making a difference as have Tall Timbers' efforts over the years. Tall Timbers has also developed a broad focus toward preservation, so the question really is do we need a global organization whose primary focus is fire management activism? I started up a Facebook page called The Association of Fire Management Activists to begin to address this issue for those who might be interested. [99]

186

If Dad and Roy had another lifetime you can bet they would have had a very major impact using Tall Timbers in an activist role to continue to pressure the large public land management bureaucracies, and I bet catastrophic fire would be on the decrease not just in the US but around the world as well. I have been impressed by what just one person or several good people can do to change things for the better, but they have to be very dedicated and driven, something you don't often see with people wanting to protect their paychecks.

We have these academics, bureaucrats and politicians with their private get along entourages feeding on taxpayer funded environmental pork projects where nobody wants to rock the boat or risk losing their funding by telling it like it is. There has to be people like me with nothing to lose to stand up and tell things like it is!!!

My friend Jack Rudloe, whom I have known most of my life and an early marine ecologist and environmental writer, is maybe even one of the founders of marine ecology. (Note as stated in the dedication, I am talking about the incompetent, corrupt academic, bureaucrat and politician. It is not my intention to paint all with the same brush. ☺)

He is completely disgusted by both private developers and environmentalists alike, working hand in hand closely with politicians on self-serving taxpayer funded environmental boondoggles. Of course nobody wants to rock the boat when they are all feeding from the same trough! If only there were more people willing to stand up and fight with their own money and time for the public good.

Such an activist organization could take activist actions that nobody else wants, or is afraid to take on, like a publicity operation mirroring the Smokey the Bear propaganda operation that has done so much damage to public consciousness. I noticed in Crawfordville, Florida that the old Smokey the Bear was still alive and well even in my own neighborhood. On the Crawfordville Highway about 25 miles south of Tallahassee, Florida close on the right heading south for everybody to see as they drive by, is Smokey with a sign saying, "Prevent Forest Fires". There was plenty of smoke from prescribed fires in the area from the prescribed fire burning of the St. Marks Refuge and Apalachicola National Forest, but there was old Smokey still standing his ground propagandizing the public.

My friend Jack and I got to talking about this problem and Jack came up with the brilliant idea of mirroring Smokey with a cartoon character Danny Duff. The idea behind Danny Duff and his supporting comic characters would to mirror and attempt to counter the very effective US Forest Service Smokey the Bear propaganda operation that has deceived the public into believing that all fire in nature is bad and destructive.

Danny Duff - If you recall the Straw man in in the Wizard of Oz was deathly afraid of fire, but in this case the Danny Duff character is an unstable, emotional, chaotic, dramatic character who loves catastrophic fire, the hotter and more destructive the better. He is drawn as a variation of the Straw man, but is a bundle of debris, pine straw, leaves and sticks.

Everywhere he goes he is leaking leaves and debris. He is to be an egotistical, rotund, dramatic and funny character lacking brains, but not mean spirited, who from time to time bursts into flames causing catastrophic wildfire. He is a good friend of Smokey the Bear and can't stand the Fire Fairy because she keeps the forest clean and neat.

Fire Fairy – The Fire Fairy is virtuous female character who is friends with all the forest and grassland animals especially the endangered Spotted Owl, the Desert Tortoise etc. She flits about lighting light fire in light fire ecosystems with her fire wand, creating diverse plant and animal habitats and ecosystems and protects them by sweeping the forest clean of debris buildup on a regular basis.

It would be nice if we could find an artist to help us create a comic strip along these lines to go viral and expose and poke fun at those who claim to be protecting our public and private forests and grasslands, but who are actually creating the conditions for the destruction of these ecosystems by catastrophic wildfire through accelerating fire suppression.

One enduring cartoon might be with Danny Duff with his arm around Smokey, his best friend, ☺ thanking him for putting out all those nuance light fires that the Fire Fairy and her animal friends have been using to sweep the forest clean. How can Danny Duff possibly start catastrophic fires without Smokey's help to build up forest fuel accumulations to a catastrophic level?

My father, when he organized Tall Timbers, pulled together diverse people from around the world who knew the importance of fire in nature in the Fire Conferences and with the organization itself. Now I think it is time to try to do this same thing and bring pro-fire activists together to create a more cohesive front to eliminate once and for all these unnatural catastrophic fires. I think it should also be focused on going beyond stopping man caused catastrophic fire to rebuilding diversity in light fire ecosystems that has been lost since the native peoples around the world were decimated during the European conquests.

Like Tall Timbers, it should be in a large part privately funded, so that in a heated political battle an institution being held accountable cannot attack its funding to make it back off. Tall Timbers in the early years was protected by plantation funding not easily attacked. A new organization could be funded by insurance companies, environmentalist organizations, individuals and private property owners who have the most to lose from continued man caused catastrophic fire. We fight fire with fire in the political arena as well as in ecosystems. This is because wrongdoers fight back to protect their special interests. ☺ People that throw up their hands and say, "why can't people just get along and stop bickering" just don't have a clue as to what it takes to combat predation from self-serving interests.

Multi-billion Dollar Wildfire Mitigation Program

In the introduction to this book, I discussed briefly Homeland Security's concern that terrorists could use wildfire as a weapon against the United States and other countries. One wonders what would happen if terrorists were able to make a significant wildfire attack against even one nation in the world using wildfire under extreme weather conditions. Would this wildfire problem be finally and quickly addressed and resolved in the developed countries like the United States and Australia?

Could huge suffering and loss of life force countries to take immediate action to drastically reduce fuel loads in light fire ecosystems for national security reasons? What if suddenly tens of billions of dollars were available to eliminate most of the wildfire danger? How could these funds be most efficiently spent in the western United States and Australia to resolve the wildfire problem along with revitalizing and restoring healthy light fire ecosystems?

My guess that governments would do as they have done before in cases of national emergency, create an agency like the WPA. Such a national agency would oversee and implement wildfire fuel reduction on a massive scale with controlled burning and mechanical means. It would override any special interest, environmental, legal, or other objections to get the job done. Therefore, this is something environmental groups should consider and think about and plan for now, so as to ensure that this debris removal would be as ecological friendly as possible, if this terrorist scenario should unfold.

Urban planners and public land management bureaucracies should also be thinking about how such a program would be implemented efficiently and with consideration for healthy fire ecosystems. The last thing our mismanaged national forests and grasslands need is even more misguided and ecologically unfriendly actions taken by humans.

As wildfire mitigation proceeds, civil liberty issues will surely arise especially among private property owners in the urban-forest interface. Landowners could be forced in short order to build wildfire defensive zones with or without help from governments, around homes and property, prior to very large scale controlled burning on public and private lands.

Public and private land managers need to be thinking and planning as to how to use prescribed fire on a massive scale involving tens of millions even hundreds of millions of acres. How can prescribed fire training be rapidly ramped up and the few well trained experts in controlled burning and fire ecology be best found utilized in this kind of Manhattan Project? How are we going to mitigate extensive wildfire destruction by not well trained and incompetent fire managers?

What if your agency suddenly had to figure out how to control burn a million acres of your public forest or grassland in one fire season? Would you be prepared with a plan in place to coordinate with Homeland Security? Maybe you even now could get some funding for some initial planning from Homeland Security.

I think right now with the Nature Conservancy taking the lead in very limited experiments in fuel reduction, they should get funding to do some Homeland Security large scale planning. The same should be considered for other NGOs like Tall Timbers. Homeland Security should be thinking about funding long range planning for government agencies and NGOs alike.

I have written this book not only to educate and inform the public, but also as an activist weapon against those responsible for the yearly destruction of millions of acres of light fire ecosystems. Like my father, I am not into this for money, I am it for both humanity and the environment, so I intend to make the book free to read on the Internet to reach the largest number of the general public and environmental activists possible. I will try to reach those who really want to make a difference by fighting for what is right and the common good, rather than for their own self-enrichment.

I have lived all my life very frugally so my life has been my own and I can concentrate on the common good and self-development rather than concentrate on materialistic pursuits that are killing off both nature and humanity! Leaders need to lead by example in order to have moral authority. Hypocritical environmentalists, just like hypocritical everybody else, have to make the switch to downsizing their personal lifestyle and reproduction. We all need to concentrate on our own personal self-development and the common good else humanity and nature's suffering will only increase catastrophically. It won't be just be by catastrophic wildfire that defeats us either! Few people really want to focus on the elephant in the room that is really driving global environmental destruction, that of over-population and over-consumption.

CHAPTER SIX

THE ELEPHANT IN THE ROOM

I suppose the question is are we making any headway toward solving the problem of getting light fire back into light fire ecosystems and putting an end to man caused catastrophic fire in these ecosystems? Some would say yes and others would say we are losing the battle, depending on the perspective they prefer, be as an optimist or a pessimist.

On the one hand, our collective knowledge on fire and fire management has grown by leaps and bounds from what it was in the 1950s setting up infrastructure that could be used to eventually get a handle on the problem of man caused catastrophic fire. Yet on the other hand, millions of acres of light fire ecosystems are being devastated every year by man caused catastrophic fire with the numbers continuing to rise rather than decline year to year.

It seems to me that what we have with the catastrophic fire problem fits into an overall fighting retreat to preserve land from man's destructive behaviors in a large part driven by overpopulation and overconsumption that continue to rise, putting more and more pressure on nature and man himself. There seems to be reluctance, even denial, on the part of the public and the environmental movement to address these fundamental factors, this overall context and instead play the blame game, or generally get distracted by peripheral issues.

The focus within the environmental movement and development industry seems to be to play distraction games blaming each other, while together trying to nibble around the edges of the elephant in the room trying to solve issues that are more symptoms of overpopulation and overconsumption than causes in themselves. Developers are like drug dealers, if there is a demand there will always be developers to fulfill that demand and lobby for even more demand out of self-interest. What good does it do to demonize developers as this distracts from real causes? Producers simply work to supply consumer demand. As every doctor knows treating symptoms, or blaming doctors or patients can lead to the continuance of the disease and the death of the patient!

Because mankind as a whole is reluctant to limit reproduction and to settle for low consumption lifestyles on an individual and collective basis, our global environmental and social problems continue to worsen. This includes the loss of wildland to development and populations retreating into the country making it more and more difficult to prescribe burn around all the development hazards. So what are we to do? Give up and roll over and let nature take its course through disease and war as with other animals, or do we take our own destiny into our own hands as a rational thinking species? Or perhaps the reality will be a combination of both.

As a species, most of us think having children is a God given right, except perhaps in China where the people there have been forced to limit their reproduction for the common good. I am very much opposed to giving up my individual liberty, but if by exercising that liberty I am destroying humanity and the environment, should I not rethink my actions and make them more in accord with common good, the very environment into which my children are being born? Obviously it would be much better for you and me to limit our rights, than have society limit them for us, but are any of you willing to do that? Me I live on about 500 dollars a month and have had no children or grandchildren, how about you?

The cells that form our bodies were once free swimming organisms in the ocean, but they gave up their liberty of free movement to combine together to form colonies of organisms that had a competitive advantage to some degree over smaller free swimming single celled organisms. This struggle between these two competitive strategies is still ongoing as our bodies fight off viruses and bacterial diseases while cooperating with other micro-organisms as in our guts.

It has been my contention that population pressures and evolution within humanity are forcing for competitive reasons some single individuals to give up liberties to combine into a more advanced multi-individual organisms. In return for liberty, the individuals in this greater humanity trade some of that liberty for an equalitarian system where all contribute to the welfare of the whole, and all reap the benefits of this cooperation, becoming ever more specialized as are cells in our bodies. This scaling up is an indication of living in a fractal universe I believe.

So where does this overall context leave us in our battle to save our light fire ecosystems from catastrophic fire? It seems to me that the Nature Conservancy is moving in the right direction, albeit slowly, by building partnerships between environmentalists and landowners and society as a whole. Somehow this idea has to be expanded to build global partnerships into movements to limit over-population, over-consumption and environmental destruction due to such things as unclean fuels.

Environmental Deception and Scamming of the Public

I am really wary of the global warming movement because it seems that it is elite corporate industry and banking leaders that are driving this movement, the very folks responsible for the problem in the first place. Such elite leaders or rulers benefit in creating a problem as in the fossil fuels industry, and then turn around to present solutions to the problem they

195

created at the expense of the rest of us. This is a tried and true manipulative technique, used throughout history by the upper classes against the lower classes of society.

Real partnerships are critical toward solving this catastrophic fire problem, but it is also critical to be really careful that the ones we are partnering with, are sincere and competent, and are not scamming us with environmental or development boondoggles. So many of our politicians today have these environmental and development entourages that actually work together to scam the taxpayers, and deprive real competent environmental projects of funds.

No matter if the taxpayer funded project is for fire suppression or fire management, we must look closely and review the project carefully to see that if the project can really deliver on its promises, or if it's just another scam of the taxpayer and the public good. I believe the really good projects are not being funded by just the taxpayer, but by private donations as well.

Private donors take how their money is used much more seriously than the taxpayer who has little understanding or control over his or her money. Still, we have to study what are the agendas and motivations of the private donors so that they are not deceiving us as to their intentions to serve the public good rather than themselves.

We hear all this talk about environmental awareness, but what about deception awareness? What about all this scamming and corruption in politics, environmental and development communities? How are we going to get a handle on this? Are we going to let the usual incompetent people gain control over the fire management community and waste what little resources that are available? Or even worse, let prescribed fire get out of control to set off even more severe wildfire leading to even more backlash against controlled fire?

Are we going to let the fire suppression industry continue the way they are going, sucking up all the funding insanely making the fuel load problem even worse, leading to even more catastrophic fire down the road? Is this another case of one or more industries helping to create a problem out of self-interest only to turn around offering ineffective solutions to further self-interest?

How are we going to get smarter competent people into government and private leadership positions to change a fire suppression culture to a fire management culture? How much more loss of life, property and ecological destruction is it going to take to move from fire suppression to overall fire management in any meaningful way? Is fire management still only a subculture embedded within an overall fire suppression culture?

Information or Psychological Warfare, the Assault on Public Fire Consciousness

We have to collectively face the fact that the consciousness of global populations has been under a global propaganda assault in regards to fire for over 100 years through government agencies like the U.S. Forest Service. Smokey the Bear became the mouthpiece of this assault that really began in earnest in the 1920s. The slogan, "Only You Can Prevent Forest Fires" assaulted the public mind everywhere it was used and spread from the United States to the rest of the world. It was a very well taxpayer funded big lie that was and still is being supported by a context of constantly repeated little lies.

A counter assault to free the public mind boils down to public education and the reason I and others write books and articles on fire management. I can only take a small bite out of this problem, but if enough of us are willing to make the personal sacrifices necessary, then we begin to have more and more influence over resolving this fire suppression problem. This is really the crux of the problem, we got to have Smokey the Bear crying out everywhere before the public saying, "Only you can use fire wisely" rather than, "Only you can prevent wildfire" or worse, " Only you can prevent forest fires." It's been like pulling teeth over decades to get Smokey just from "Only you can prevent forest fires" to "Only you can prevent wildfire". Let's face it, Smokey has yet to really be enthusiastically reformed, only giving as much ground as is absolutely necessary.

Words and language are very powerful, they both can be used to heal, or they can be used as weapons to enslave by influencing thinking, emotion and culture. We know a society, a culture, both by the words it uses and the way it acts, not necessarily by what it says as to its intentions and motivations.

So ultimately our struggle for prescribed fire is a mental, emotional, and a spiritual one. It is a struggle against the false memes embedded within individuals and society. False memes are obstacles to enlightenment and advancement on all fronts. These false memes have to be replaced with truer ones in our fight for public education, truth, justice and light fire in our forests and grasslands, even as we struggle mightily with our actions to right the wrongs of the past.

One of the most prevalent and destructive false memes that infect us all is, "Do as I say, not as I do." As with the rest of our society, this meme infects the environmental movement where the very people with the most resources and education living middle class or upper class lifestyles, rave and rant over the destruction of the environment that they themselves in a large part are driving through excessive reproduction and overconsumption. Of course the lower classes are involved, but leaders mostly come out of the middle and upper classes and it is they by their actions and deeds that set the example for the poor and the less educated and control to a large extent the lives of the poor through control of resources and propaganda.

Mahatma Gandhi could be considered one of the most sincere and non-hypocritical leaders of humanity in the 20th century. His leadership has been of tremendous benefit to humanity inspiring others like the leaders of the Civil Rights Movement toward non-violence by himself practicing non-violence in the face of great violence. What often gets overlooked is the fact that Gandhi had almost no possessions as did Christ and Buddha even earlier in humanity's past. How many middle and upper class environmentalists are going to really downsize their own lifestyles and quit having children before forcing others to do the same?

So we see in the overall environmental movement context and in the fire management movement sub-context, a very serious moral issue needing to be aired and resolved, if we are to speak with real moral authority to the rest of mankind. It's the middle and the upper classes that are benefiting the most materialistically by hording, flaunting and inefficiently consuming scarce planetary resources. Those that control resources control people and so they are the leaders or rulers of society and so have the greater responsibility toward transforming society for the better, both by action and by example.

The upshot is that our wildlands are rapidly decreasing as population and consumption pressures escalate, and combined with ignorance of fire, is leading to unnecessary destruction of the remaining light fire ecosystems to catastrophic fire. Making things even worse, are the pressures to build homes and industry in rural and wildland areas making it even more difficult to use prescribed fire. We can begin to adapt to work with homeowners to create defensive zones around houses as the Nature Conservancy is doing, but is it any more than a fighting retreat against this overall cultural context of over-population and over-consumption? Is this the best we can do?

Maybe the best we can do to try to preserve and manage our disappearing and transforming landscapes is to try our level best to educate the public, not propagandize them. We got to strip away the chains that Smokey the Bear has laid on society in regards to the importance of fire in light fire ecosystems. We got to free our minds of the ignorance of fires natural role in earth's ecosystems. We have to win the information war for the hearts and minds of global populations to get fire back into these ecosystems as quickly as possible. We all got to become activists and push forward to overcome the obstacles not just on the land and sea, but in our own minds and bodies. We have got to become better people to adequately solve the problems associated with fire and with humanity as a whole.

This requires accepting simple lifestyles that promote free time to reflect and improve on our actions to become such better people. Highly stressed populations are not healthy populations, nor are they healthy for the environment. We got to find the time to stand back and gain perspective on our lives. Living life is like painting a painting. One gets up close and works on the details, but then stands back to gain perspective to make sure that the details don't get out of proportion and run together and become like modern art.

Materialism is rampant today and is in a large part responsible for human and nature's woes. Our gods have become materialistic gods of pleasure and extravagance, shortsighted in the extreme. We worship materialism, the superficial aspects of life that which we can easily see and ignore the fundamental often hidden or unseen forces that shape our lives for better or worse. It's like we can see the trees, but the ecology of the forest ecosystem is beyond our comprehension. Because of this, we as a species suffer, physically, emotionally and mentally, and we are no happier no matter how hard we search for materialistic solutions always just beyond our reach like the illusionary pot of gold at the end of the rainbow.

Of what use are money, power, resources and material things in their own right if they are not used within a greater moral and ethical context of immaterial quantum reality in which material holographic reality is embedded. Of what use is a car if the driver is out of his or mind? Of what use are all these material things and people we are bringing into the world, if we have created an environmental pigsty for ourselves and our children and remain an unhappy stressed out humanity?

Young people, you must take the moral authority, become activists, take the ball and run with it for the rest of your lives, to not only end this fire suppression nightmare, but to transform society for the benefit of both humanity and nature. Remember also that your own self development is part of your mission in life. Self-development is just as important as external development. You must not lose sight of this as you proceed in life within a short-sighted society that does not put enough value on individual inner betterment and insight. We live in a society that will pull you down into the quagmire of selfishness and materialism if you do not remain strong and vigilant all your life.

When you are down and stressed out, make the time to go out into or find a little remaining wildland and care for it as you do your own family, to maintain balance and inner peace. Live in the eye of the hurricane where the sun shines and the wind is calm. Live simply, quietly and effectively, and when you have your center, then move in and out of society to transform society for the better. My father told me that the Longleaf Pine forest on Greenwood was his cathedral and I read that many of his associates have said basically the same thing.

When you are old, don't forget to give back, to write down for posterity what you have learned, so others can build on your good works just as you have built your life on the good works of others. Our lifetimes are short and in order to continue to advance and evolve, we must develop better and better ways to transfer good credible knowledge to our younger generation. Do this so that the younger generation can keep moving humanity forward, building a humanity that can live in harmony with fire and nature even as we travel to the stars.

OTHER BOOKS BY THE AUTHOR

UFOs Exopolitics and the New World Disorder is the first book in a series of books written by Ed Komarek and is free on its website. It can be ordered in print and on Kindle from Amazon. http://authors.exopaedia.org/edkomarek/index.html

Fire in Nature, A Fire Activist's guide is the second in a series of books being written by Ed Komarek for the younger generation. The book is free on its website, or can be ordered in print and Kindle on Amazon. http://fireinnature.weebly.com/

The Long Hard Road to Enlightenment is the third book in this series and is in the process of being written. It will be in the style of the previous two books including many footnoted links to credible source material.

The New World Order Disorder is in the conceptual phase and will be about international politics as its name implies.

Return to Alaska is in the conceptual phase and will be about growing up to become an independent, self-sufficient, frugal sovereign individual.

Ed Komarek has been mentored by many people in his long life by those that went out of their way to help him in his life's journey and mission. In appreciation of that, he feels he should return the favor to society and the younger generation. He can be reached at edkomarek@yahoo.com or on Facebook message.

Footnotes
Supporting document hyperlink index

[1] Fire in Nature, A Fire Activists Guide:
http://fireinnature.weebly.com/
[2] UFOs Exopolitics and the New World Disorder:
http://authors.exopaedia.org/edkomarek/index.html
[3] Gulf Specimen Marine Lab:
http://www.gulfspecimen.org/
[4] Wildfire Today
http://wildfiretoday.com/tag/statistics/
[5] Headwaters Economics
http://headwaterseconomics.org/
[6] Interagency Fire Center
http://www.nifc.gov/fireInfo/fireInfo_statistics.html
[7] Tall Timbers Research Station
https://www.talltimbers.org/
[8] Homeland Security Warns of Terrorist Wildfire Attacks
http://publicintelligence.net/homeland-security-warns-of-
terrorist-wildfire-attacks/
[9] Geologic Time Scale
http://en.wikipedia.org/wiki/Geologic_time_scale
[10] The Diversification of Paleozoic fire Systems and
Fluctuations in Atmospheric Oxygen
http://www.ncbi.nlm.nih.gov/pmc/articles/PMC1544139/
[11] Ordovician Image
http://www.gambassa.com/gambassafiles/images/images/2035/
120201094923_large_v2.jpeg
[11] Landmass Image
http://geology.gsapubs.org/content/33/2/109/F1.large.jpg
[12] How Plants Helped Make the Earth Unique
http://www.livescience.com/18254-plants-extinction-climate-
change-rivers.html
[13] Silurian Landscape Image
http://img1.wikia.nocookie.net/_cb20120802164306/worldofpr
ehistory/images/b/b0/Silurian_flora.jpg

[14] Silurian Landmass Image
http://www.paleoportal.org/media/boilerplate/0/12573_period_pal_map_14_image.jpg
[15] Silurian Oxygen Levels
http://eonsepochsetc.com/Paleozoic/Silurian/silur_home.html
[16] Fossil Record of Fire
http://en.wikipedia.org/wiki/Fossil_record_of_fire
[17] Devonian
http://en.wikipedia.org/wiki/Devonian
[18] Devonian Image
http://museumvictoria.com.au/pages/17107/imagegallery/1devonian-pic-41884.jpg
[19] Landmass Image
http://www.scotese.com/images/390.jpg
[20] Silurian and Devonian
http://steurh.home.xs4all.nl/engplant/eblad1.html
[21] Newfound Fossils Reveal secrets of World's Oldest Forest
http://news.nationalgeographic.com/news/2007/04/070418-oldest-trees_2.html
[22] Carboniferous
http://en.wikipedia.org/wiki/Carboniferous
[23] Carboniferous Image
http://www.arcadiastreet.com/cgvistas/earth/02_paleozoic/images/carboniferous_firestorm_v2_800.jpg
[24] Carboniferous Landmass Image
http://www.scotese.com/images/306.jpg
[25] Permian
http://en.wikipedia.org/wiki/Permian
[26] Permian Image
http://fc00.deviantart.net/fs27/f/2008/096/6/d/Permian_fauna_from_Marocco_by_dustdevil.jpg
[27] Permian Landmass Image
http://upload.wikimedia.org/wikipedia/commons/5/56/280_Ma_plate_tectonic_reconstruction.png
[28] Triassic Image

http://images.nationalgeographic.com/wpf/media-live/photos/000/012/cache/triassic-dinosaurs_1256_600x450.jpg
[29] Triassic Landmass Image
http://upload.wikimedia.org/wikipedia/commons/e/e6/Blakey_220moll.jpg
[30] Triassic
http://en.wikipedia.org/wiki/Triassic
[31] Jurassic Image
http://upload.wikimedia.org/wikipedia/commons/1/1c/Europasaurus_holgeri_Scene_2.jpg
[32] Jurassic Landmass Image
http://palaeos.com/mesozoic/jurassic/images/MiddleJurassicMap.jpg
[33] Jurassic
http://en.wikipedia.org/wiki/Jurassic
[34] Coniferophyta
http://www2.mcdaniel.edu/Biology/botf99/gynopsperms/conifers.html
[35] Cretaceous Image
http://www.eos.ubc.ca/courses/Dist-Ed/EOSC116/images/moduleC-lesson07/Karen_Carr_Cretaceous_Coastal_Landscape.jpg
[36] Cretaceous Landmass Image
http://upload.wikimedia.org/wikipedia/commons/d/dc/Blakey_90moll.jpg
[37] Cretaceous Period
http://www.britannica.com/EBchecked/topic/142729/Cretaceous-Period
[38] Angiosperm
http://www.britannica.com/EBchecked/topic/24667/angiosperm
[39] Grass
http://en.wikipedia.org/wiki/Grass
[40] Cretaceous Wildfires and Their Impact on the Earth System

http://www.sciencedirect.com/science/article/pii/S0195667112
00016X
[41] Cenozoic Image
http://3.bp.blogspot.com/-
Y3L57hdPnUU/Thuql9_jbqI/AAAAAAAAJs/l2dP6rlRWq8/
s1600/2004_gallery_pliestocene.jpg
[42] Cenozoic Land Mass Image
http://media.web.britannica.com/eb-media/13/813-004-
10B408EB.gif
[43] The Evolution of Plants
http://steurh.home.xs4all.nl/engplant/eblad1.html
[44] Late Cenozoic Image
http://encyclopediaurantia.org/images/pra292.jpg
[45] Ardipithecus Image
https://www.sciencenews.org/sites/default/files/11686
[46] Homo Habilis Image
http://humanorigins.si.edu/evidence/human-
fossils/species/homo-habilis
[47] Human Evolution
https://en.wikipedia.org/wiki/Human_evolution
[48] Homo Erectus
http://humanorigins.si.edu/evidence/human-
fossils/species/homo-erectus
[49] Homo Erectus
http://en.wikipedia.org/wiki/Homo_erectus
[50] Control of Fire by Early Humans
http://en.wikipedia.org/wiki/Control_of_fire_by_early_humans
[51] Who Mastered Fire?
http://www.slate.com/articles/health_and_science/human_evol
ution/2012/10/who_invented_fire_when_did_people_start_coo
king_.html
[51] Evidence That Human Ancestors Used Fire One Million
Years Ago
http://www.sciencedaily.com/releases/2012/04/120402162548.
htm
[52] Early Human Fire Skills Revealed

http://news.bbc.co.uk/2/hi/science/nature/3670017.stm
[53] Stone Age Man Used Fire to Make Tools – 50,000 Years Ago
http://www.telegraph.co.uk/science/science-news/6023724/Stone-Age-man-used-fire-to-make-tools-50000-years-earlier-than-we-scientists-thought.html
[54] Holocene
http://en.wikipedia.org/wiki/Holocene
[55] Reconstructing Holocene Fire History in a Southern Appalachian Forest Using Soil Charcoal
http://www.ncbi.nlm.nih.gov/pubmed/20426326
[56] Global Palaeofire Working Group
http://gpwg.org/gpwganalyses.html
[57] Historical Ecology
http://en.wikipedia.org/wiki/Historical_ecology
[58] The World on Fire
http://www.pbs.org/wgbh/nova/fire/world.html
[59] Subsistence Hunting and Gathering
http://www.fao.org/docrep/w7540e/w7540e0i.htm
[60] The Hadza or Bushmen of Tanzania
http://www.adventureclassroom.org/cultures/bushmen.htm
[61] Aboriginal Hunting and Burning Increase Australia's Desert Biodiversity
http://news.stanford.edu/news/2010/april/martu-burning-australia-042910.html
[62] Early Human Migrations
https://en.wikipedia.org/wiki/Early_human_migrations
[63] Early Humans Settled India Before Europe, Study Suggests
http://news.nationalgeographic.com/news/2005/11/1114_05111 4_india.html
[64] Recent African Origin of Modern Humans
https://en.wikipedia.org/wiki/Recent_African_origin_of_modern_humans
[65] A Fiery History
http://www.atree.org/fiery_history

[66] The Soliga Community of Karnataka and Their Intimate Relationship With Nature
http://archive.today/TNdMy
[67] China Pilots Wildfire Detection Sensor Network
http://www.futuregov.asia/articles/2012/jan/18/china-pilots-wildfire-detection-sensor-network/
[68] Fire Landscapes of China
http://www.firechief.com/wf-tactics/exploring-forest-management-fire-suppression-and-environmental-conservation-china
[69] Cenozoic Plate Reconstruction of Southeast Asia
http://www.sciencedirect.com/science/article/pii/0040195195000232
[70] Cenozoic Era
http://www.britannica.com/EBchecked/topic/101936/Cenozoic-Era
[70] Neanderthals Were Nifty At Controlling Fire
http://www.sciencedaily.com/releases/2011/03/110314152917.htm
[71] Europe's First Farmers Were Immigrants: Replaced Their Stone Age Hunter-gatherer Forerunners
http://www.sciencedaily.com/releases/2009/09/090903163902.htm
[72] Fertile Crescent
http://en.wikipedia.org/wiki/Fertile_Crescent
[73] World on Fire
http://www.pbs.org/wgbh/nova/fire/world.html
[74] Cenozoic
http://en.wikipedia.org/wiki/Cenozoic
[75] Native American Use of Fire
http://en.wikipedia.org/wiki/Native_American_use_of_fire
[76] Maize
http://en.wikipedia.org/wiki/Maize
[77] Great Plains
http://en.wikipedia.org/wiki/Great_Plains
[78] History of the Giant Sequoia

http://www.monumentaltrees.com/en/trees/giantsequoia/history/

[79] Fires in Alaska
http://www.alaskacenters.gov/fires-in-alaska.cfm
[80] Grasslands National Park
http://www.pc.gc.ca/pn-np/sk/grasslands/natcul/natcul1.aspx
[81] Grasslands National Park – Prescribed Fires
http://www.pc.gc.ca/pn-np/sk/grasslands/natcul/natcul8.aspx
[82] The Use of Prescribed Fire in the Management of Canada's Forested Lands
http://cfs.nrcan.gc.ca/publications/?id=3162
[83] World on Fire
http://www.pbs.org/wgbh/nova/fire/world.html
[84] Prescribed Fire Research in the Chaco Region
http://www.fire.uni-freiburg.de/iffn/country/ra/ra_5.html
[85] Burning, Fire Prevention and Landscape Productions Among the Pemon
http://www.ncbi.nlm.nih.gov/pubmed/23246908
[86] Major Habitats & Plant Communities of the Barnegat Bay Ecosystem
http://bbp.ocean.edu/pages/138.asp
[87] Birdsong Nature Center
http://www.birdsongnaturecenter.org/
[88] Forestry
http://en.wikipedia.org/wiki/Forestry
[89] Wildland Fire Lessons Learned Center
http://www.wildfirelessons.net/Communities/Resources/ViewIncident/?DocumentKey=519f8b04-1c09-4af3-a835-1256ab1169a9
[90] Chapman Called 'Father of Controlled Burning'
http://www.thepineywoods.com/ChapmanNov08.htm
[91] Tall Timbers E. V. Komarek Fire Ecology Database
http://www.frames.gov/rcs/ttrs/11000/11859.html
[92] National Interagency Prescribed Fire Training Center
http://www.fws.gov/fire/pftc/
[93] Wildfire Experts Call for More Controlled Burns

http://www.usatoday.com/story/news/nation/2013/07/07/fire-experts-call-for-more-controlled-burns-to-stem-wildfires/2495873/

[94] Fire Science, Fire Culture
http://firehistory.asu.edu/fire-science-fire-culture/

[95] Wildfire Experts Call for More Controlled Burns
http://www.usatoday.com/story/news/nation/2013/07/07/fire-experts-call-for-more-controlled-burns-to-stem-wildfires/2495873/

[96] Saving Lives and Property From Wildfire
http://www.firewise.org/?sso=0

[97] Nature Conservancy Offers Alternative to Aggressive Fire Suppression
http://www.denverpost.com/ci_23366318/nature-conservancy-offers-alternative-aggressive-fire-suppression

[98] Fire and Conservation: What We Do
http://www.nature.org/ourinitiatives/habitats/forests/howwework/integrated-fire-management.xml

[99] Association of Fire Management Activists
https://www.facebook.com/groups/234543706747903/#_=_